Statistics and Computing

Series Editors:
J. Chambers
W. Eddy
W. Härdle
S. Sheather
L. Tierney

Springer

New York
Berlin
Heidelberg
Barcelona
Hong Kong
London
Milan
Paris
Singapore
Tokyo

Statistics and Computing

Gentle: Numerical Linear Algebra for Applications in Statistics.
Gentle: Random Number Generation and Monte Carlo Methods.
Härdle/Klinke/Turlach: XploRe: An Interactive Statistical Computing Environment.
Krause/Olson: The Basics of S and S-PLUS.
Lange: Numerical Analysis for Statisticians.
Loader: Local Regression and Likelihood.
Ó Ruanaidh/Fitzgerald: Numerical Bayesian Methods Applied to Signal Processing.
Pannatier: VARIOWIN: Software for Spatial Data Analysis in 2D.
Venables/Ripley: Modern Applied Statistics with S-PLUS, 3rd edition.
Wilkinson: The Grammar of Graphics

Clive Loader

Local Regression
and Likelihood

With 68 Figures

Springer

Clive Loader
Lucent Technologies
600 Mountain Ave., Rm. 2C-279
Murray Hill, NJ 07974-0636
USA
clive@research.bell-labs.com

Series Editors:

J. Chambers
Bell Labs, Lucent Technologies
600 Mountain Ave.
Murray Hill, NJ 07974
USA

W. Eddy
Department of Statistics
Carnegie Mellon University
Pittsburgh, PA 15213
USA

W. Härdle
Institut für Statistik und
  Ökonometrie
Humboldt-Universität zu Berlin
Spandauer Str. 1
D-10178 Berlin
Germany

S. Sheather
Australian Graduate School
  of Management
University of New South Wales
Sydney NSW 2052
Australia

L. Tierney
School of Statistics
University of Minnesota
Vincent Hall
Minneapolis, MN 55455
USA

Library of Congress Cataloging-in-Publication Data
Loader, Clive.
Local regression and likelihood / Clive Loader.
    p.   cm. — (Statistics and computing)
    Includes bibliographical references and index.
    ISBN 0-387-98775-4 (alk. paper)
    1. Regression analysis. 2. Estimation theory. I. Title.
II. Series.
QA276.8.L6   1999
519.5'36—dc21   99-14732

Printed on acid-free paper.   # 41002673

Lucent Technologies
Bell Labs Innovations

Production managed by Robert Bruni; manufacturing supervised by Jacqui Ashri.
Photocomposed pages prepared from the author's LATEX files.
Printed and bound by Maple-Vail Book Manufacturing Group, York, PA.
Printed in the United States of America.

9 8 7 6 5 4 3 2 1

ISBN 0-387-98775-4 Springer-Verlag New York Berlin Heidelberg   SPIN 10712641

Preface

This book, and the associated software, have grown out of the author's work in the field of local regression over the past several years. The book is designed to be useful for both theoretical work and in applications. Most chapters contain distinct sections introducing methodology, computing and practice, and theoretical results. The methodological and practice sections should be accessible to readers with a sound background in statistical methods and in particular regression, for example at the level of Draper and Smith (1981). The theoretical sections require a greater understanding of calculus, matrix algebra and real analysis, generally at the level found in advanced undergraduate courses. Applications are given from a wide variety of fields, ranging from actuarial science to sports.

The extent, and relevance, of early work in smoothing is not widely appreciated, even within the research community. Chapter 1 attempts to redress the problem. Many ideas that are central to modern work on smoothing: local polynomials, the bias-variance trade-off, equivalent kernels, likelihood models and optimality results can be found in literature dating to the late nineteenth and early twentieth centuries.

The core methodology of this book appears in Chapters 2 through 5. These chapters introduce the local regression method in univariate and multivariate settings, and extensions to local likelihood and density estimation. Basic theoretical results and diagnostic tools such as cross validation are introduced along the way. Examples illustrate the implementation of the methods using the LOCFIT software.

The remaining chapters discuss a variety of applications and advanced topics: classification, survival data, bandwidth selection issues, computation and asymptotic theory. Largely, these chapters are independent of each other, so the reader can pick those of most interest.

Most chapters include a short set of exercises. These include theoretical results; details of proofs; extensions of the methodology; some data analysis examples and a few research problems. But the real test for the methods is whether they provide useful answers in applications. The best exercise for every chapter is to find datasets of interest, and try the methods out!

The literature on mathematical aspects of smoothing is extensive, and coverage is necessarily selective. I attempt to present results that are of most direct practical relevance. For example, theoretical motivation for standard error approximations and confidence bands is important; the reader should eventually want to know precisely *what* the error estimates represent, rather than simply asuming software reports the right answers (this applies to any model and software; not just local regression and LOC-FIT!). On the other hand, asymptotic methods for boundary correction receive no coverage, since local regression provides a simpler, more intuitive and more general approach to achieve the same result.

Along with the theory, we also attempt to introduce understanding of the results, along with their relevance. Examples of this include the discussion of non-identifiability of derivatives (Section 6.1) and the problem of bias estimation for confidence bands and bandwidth selectors (Chapters 9 and 10).

Software

Local fitting should provide a practical tool to help analyse data. This requires software, and an integral part of this book is LOCFIT. This can be run either as a library within R, S and S-Plus, or as a stand-alone application. Versions of the software for both Windows and UNIX systems can be downloaded from the LOCFIT web page,

http://cm.bell-labs.com/stat/project/locfit/

Installation instructions for current versions of LOCFIT and S-Plus are provided in the appendices; updates for future versions of S-Plus will be posted on the web pages.

The examples in this book use LOCFIT in S (or S-Plus), which will be of use to many readers given the widespread availability of S within the statistics community. For readers without access to S, the recommended alternative is to use LOCFIT with the R language, which is freely available and has a syntax very similar to S. There is also a stand-alone version, C-LOCFIT, with its own interface and data management facilities. The interface allows access to almost all the facilities of LOCFIT's S interface, and a few additional features. An on-line **example** facility allows the user to obtain C-LOCFIT code for most of the examples in this book.

It should also be noted this book is not an introduction to S. The reader using LOCFIT with S should already be familiar with S fundamentals, such as reading and manipulating data and initializing graphics devices. Books such as Krause and Olson (1997), Spector (1994) and Venables and Ripley (1997) cover this material, and much more.

Acknowledgements

Acknowledgements are many. Foremost, Bill Cleveland introduced me to the field of local fitting, and his influence will be seen in numerous places. Vladimir Katkovnik is thanked for helpful ideas and suggestions, and for providing a copy of his 1985 book.

LOCFIT has been distributed, in various forms, over the internet for several years, and feedback from numerous users has resulted in significant improvements. Kurt Hornik, David James, Brian Ripley, Dan Serachitopol and others have ported LOCFIT to various operating systems and versions of R and S-Plus.

This book was used as the basis for a graduate course at Rutgers University in Spring 1998, and I thank Yehuda Vardi for the opportunity to teach the course, as well as the students for not complaining too loudly about the drafts inflicted upon them.

Of course, writing this book and software required a flawlessly working computer system, and my system administrator Daisy Nguyen recieves the highest marks in this respect!

Many of my programming sources also deserve mention. Horspool (1986) has been my usual reference for C programming. John Chambers provided S, and patiently handled my bug reports (which usually turned out as LOCFIT bugs; not S!). Curtin University is an excellent online source for X programming (http://www.cs.curtin.edu.au/units/).

C. Loader
Murray Hill, NJ

Contents

1
The Origins of Local Regression

The problem of smoothing sequences of observations is important in many branches of science. In this chapter the smoothing problem is introduced by reviewing early work, leading up to the development of local regression methods.

Early works using local polynomials include an Italian meteorologist Schiaparelli (1866), an American mathematician De Forest (1873) and a Danish actuary Gram (1879) (Gram is most famous for developing the Gram-Schmidt procedure for orthogonalizing vectors). The contributions of these authors are reviewed by Seal (1981), Stigler (1978) and Hoem (1983) respectively.

This chapter reviews development of smoothing methods and local regression in actuarial science in the late nineteenth and early twentieth centuries. While some of the ideas had earlier precedents, the actuarial literature is notable both for the extensive development and widespread application of procedures. The work also forms a nice foundation for this book; many of the ideas are used repeatedly in later chapters.

1.1 The Problem of Graduation

Figure 1.1 displays a dataset taken from Spencer (1904). The dataset consists of human mortality rates; the x-axis represents the age and the y-axis the mortality rate. Such data would be used by a life insurance company to determine premiums.

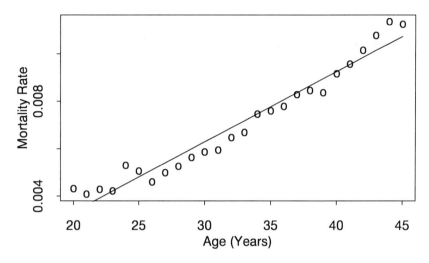

FIGURE 1.1. Mortality rates and a least squares fit.

Not surprisingly, the plot shows the mortality rate increases with age, although some noise is present. To remove noise, a straight line can be fitted by least squares regression. This captures the main increasing trend of the data.

However, the least squares line is not a perfect fit. In particular, nearly all the data points between ages 25 and 40 lie below the line. If the straight line is to set premiums, this age group would be overcharged, effectively subsidizing other age groups. While the difference is small, it could be quite significant when taken over a large number of potential customers. A competing company that recognizes the subsidy could profit by targeting the 25 to 40 age group with lower premiums and ignoring other age groups.

We need a more sophisticated fit than a straight line. Since the causes of human mortality are quite complex, it is difficult to derive on theoretical grounds a reasonable model for the curve. Instead, the data should guide the form of the fit. This leads to the problem of graduation:[1] adjust the mortality rates in Figure 1.1 so that the graduated values of the series capture all the main trends in the data, but without the random noise.

1.1.1 Graduation Using Summation Formulae

Summation formulae are used to provide graduated values in terms of simple arithmetic operations, such as moving averages. One such rule is given by Spencer (1904):

---

[1]Sheppard (1914a) reports "I use the word (graduation) under protest".

1. Perform a 5-point moving sum of the series, weighting the observations using the vector $(-3, 3, 4, 3, -3)$.

2. On the resulting series, perform three unweighted moving sums, of length 5, 4 and 4 respectively.

3. Divide the result by 320.

This rule is known as Spencer's 15-point rule, since (as will be shown later) the graduated value $\hat{y}_j$ depends on the sequence of 15 observations $y_{j-7}, \ldots, y_{j+7}$. A compact notation is

$$\hat{y}_j = \frac{S_{5,4,4}}{5 \cdot 4 \cdot 4 \cdot 4} \left(-3y_{j-2} + 3y_{j-1} + 4y_j + 3y_{j+1} - 3y_{j+2} \right). \qquad (1.1)$$

Rules such as this can be computed by a sequence of straightforward arithmetic operations. In fact, the first weighted sum was split into several steps by Spencer, since

$$-3y_{j-2} + 3y_{j-1} + 4y_j + 3y_{j+1} - 3y_{j+2}$$
$$= \; y_j + 3 \left((y_{j-1} + y_j + y_{j+1}) - (y_{j-2} + y_{j+2}) \right).$$

In its raw form, Spencer's rule has a boundary problem: Graduated values are not provided for the first seven and last seven points in the series. The usual solution to this boundary problem in the early literature was to perform some ad hoc extrapolations of the series. For the moment, we adopt the simplest possibility, replicating the first and last values to an additional seven observations.

An application of Spencer's 15-point rule to the mortality data is shown in Figure 1.2. This fit appears much better than the least squares fit in Figure 1.1; the overestimation in the middle years has largely disappeared. Moreover, roughness apparent in the raw data has been smoothed out and the fitted curve is monotone increasing.

On the other hand, the graduation in Figure 1.2 shows some amount of noise, in the form of wiggles that are probably more attributable to random variation than real features. This suggests using a graduation rule that does more smoothing. A 21-point graduation rule, also due to Spencer, is

$$\hat{y}_j = \frac{S_{7,5,5}}{350} \left(-y_{j-3} + y_{j-1} + 2y_j + y_{j+1} - y_{j+3} \right).$$

Applying this rule to the mortality data produces the fit in the bottom panel of Figure 1.2. Increasing the amount of smoothing largely smooths out the spurious wiggles, although the weakness of the simplistic treatment of boundaries begins to show on the right.

What are some properties of these graduation rules? Graduation rules were commonly expressed using the difference operator:

$$\nabla y_i = y_{i+1/2} - y_{i-1/2}.$$

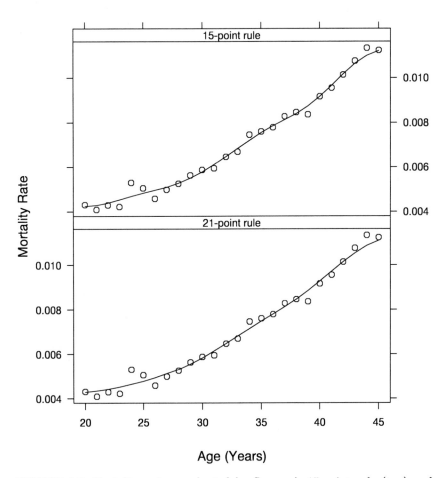

FIGURE 1.2. Mortality rates graduated by Spencer's 15-point rule (top) and 21-point rule (bottom).

The $\pm 1/2$ in the subscripts is for symmetry; if y_i is defined for integers i, then ∇y_i is defined on the half-integers $i = 1.5, 2.5, \dots$ The second differences are

$$
\begin{aligned}
\nabla^2 y_i &= \nabla(\nabla y_i) \\
&= \nabla y_{i+1/2} - \nabla y_{i-1/2} \\
&= (y_{i+1} - y_i) - (y_i - y_{i-1}) \\
&= y_{i+1} - 2y_i + y_{i-1}.
\end{aligned}
$$

Linear operators, such as a moving average, can be written in terms of the difference operator. The 3-point moving average is

$$\frac{y_{i-1} + y_i + y_{i+1}}{3} = y_i + \frac{1}{3}(y_{i-1} - 2y_i + y_{i+1})$$
$$= y_i + \frac{1}{3}\nabla^2 y_i.$$

Similarly, the 5-point moving average is

$$\frac{y_{i-2} + y_{i-1} + y_i + y_{i+1} + y_{i+2}}{5} = y_i + \nabla^2 y_i + \frac{1}{5}\nabla^4 y_i.$$

A similar form for the general k-point moving average is given by the following result.

Theorem 1.1 The k-point moving average has the representation

$$\frac{S_k}{k} y_i = \left(I + \frac{k^2 - 1}{24}\nabla^2 + \frac{(k^2 - 1)(k^2 - 9)}{1920}\nabla^4 + O(\nabla^6)\right) y_i. \qquad (1.2)$$

Proof: We derive the ∇^2 term for k odd. The proof is completed in Exercise 1.1.

One can formally construct the series expansion (and hence conclude existence of an expansion like (1.2)) by beginning with an $O(\nabla^{k-1})$ term and working backwards.

To explicitly derive the ∇^2 term, let $y_i = i^2/2$, so that $\nabla^2 y_i = 1$, and all higher order differences are 0. In this case, the first two terms of (1.2) must be exact. At $i = 0$, the moving average for $y_i = i^2/2$ is

$$\frac{S_k}{k} y_0 = \frac{1}{k} \sum_{j=-(k-1)/2}^{(k-1)/2} \frac{j^2}{2} = \frac{k^2 - 1}{24} = y_0 + \frac{k^2 - 1}{24}\nabla^2 y_0.$$

$\square$

Using the result of Theorem 1.1, Spencer's rules can be written in terms of the difference operator. First, note the initial step of the 15-point rule is

$$-3y_{j-2} + 3y_{j-1} + 4y_j + 3y_{j+1} - 3y_{j+2} = 4y_j - 9\nabla^2 y_j - 3\nabla^4 y_j$$
$$= 4(I - \frac{9}{4}\nabla^2 + O(\nabla^4))y_j.$$

Since this step is followed by the three moving averages, the 15-point rule has the representation, up to $O(\nabla^4 y_j)$,

$$\hat{y}_j = (I + \frac{5^2 - 1}{24}\nabla^2)(I + \frac{4^2 - 1}{24}\nabla^2)(I + \frac{4^2 - 1}{24}\nabla^2)(I - \frac{9}{4}\nabla^2)y_j + O(\nabla^4 y_j).$$
$$(1.3)$$

Expanding this further yields

$$\hat{y}_j = y_j + O(\nabla^4 y_j). \tag{1.4}$$

In particular, the second difference term, $\nabla^2 y_i$, vanishes. This implies that Spencer's rule has a cubic reproduction property: since $\nabla^4 y_j = 0$ when y_j is a cubic polynomial, $\hat{y}_j = y_j$. This has important consequences; in particular, the rule will tend to faithfully reproduce peaks and troughs in the data. Here, we are temporarily ignoring the boundary problem.

An alternative way to see the cubic reproducing property of Spencer's formulae is through the *weight diagram*. An expansion of (1.1) gives the explicit representation

$$
\begin{aligned}
\hat{y}_j \;=\; \frac{1}{320}(& -3y_{j-7} - 6y_{j-6} - 5y_{j-5} + 3y_{j-4} + 21y_{j-3} \\
& +46y_{j-2} + 67y_{j-1} + 74y_j + 67y_{y+1} + 46y_{j+2} \\
& +21y_{j+3} + 3y_{j+4} - 5y_{j+5} - 6y_{j-6} - 3y_{j-7}).
\end{aligned}
$$

The weight diagram is the coefficient vector

$$
\begin{aligned}
\frac{1}{320}(-3 \quad &-6 \quad -5 \quad 3 \quad 21 \quad 46 \quad 67 \quad 74 \\
& 67 \quad 46 \quad 21 \quad 3 \quad -5 \quad -6 \quad -3).
\end{aligned} \tag{1.5}
$$

Let $\{l_k; k = -7, \ldots, 7\}$ be the components of the weight diagram, so $\hat{y}_j = \sum_{k=-7}^{7} l_k y_{j+k}$. Then one can verify

$$\sum_{k=-7}^{7} l_k = 1$$

$$\sum_{k=-7}^{7} k l_k = 0$$

$$\sum_{k=-7}^{7} k^2 l_k = 0$$

$$\sum_{k=-7}^{7} k^3 l_k = 0. \tag{1.6}$$

Suppose for some j and coefficients a, b, c, d the data satisfy $y_{j+k} = a + bk + ck^2 + dk^3$ for $|k| \le 7$. That is, the data lie exactly on a cubic polynomial. Then

$$\hat{y}_j = \sum_{k=-7}^{7} l_k y_{j+k} = a \sum_{k=-7}^{7} l_k + b \sum_{k=-7}^{7} k l_k + c \sum_{k=-7}^{7} k^2 l_k + d \sum_{k=-7}^{7} k^3 l_k = a.$$

That is, $\hat{y}_j = a = y_j$.

1.1.2   The Bias-Variance Trade-Off

Graduation rules with long weight diagrams result in a smoother graduated series than rules with short weight diagrams. For example, in Figure 1.2, the 21-point rule produces a smoother series than the 15-point rule. To provide guidance in choosing a graduation rule, we want a simple mathematical characterization of this property.

The observations y_j can be decomposed into two parts: $y_j = \mu_j + \epsilon_j$, where (Henderson and Sheppard 1919) μ_j is "the true value of the function which would be arrived at with sufficiently broad experience" and ϵ_j is "the error or departure from that value". A graduation rule can be written

$$\hat{y}_j = \sum l_k y_{j+k} = \sum l_k \mu_{j+k} + \sum l_k \epsilon_{j+k}.$$

Ideally, the graduation should reproduce the systematic component as closely as possible (so $\sum l_k \mu_{j+k} \approx \mu_j$) and remove as much of the error term ($\sum l_k \epsilon_{j+k} \approx 0$) as possible.

For simplicity, suppose the errors ϵ_{j+k} all have the same *probable error*, or variance, σ^2, and are uncorrelated. The probable error of the graduated values is $\sigma^2 \sum l_k^2$. The *variance reducing factor* $\sum l_k^2$ measures reduction in probable error for the graduation rule. For Spencer's 15-point rule, the variance reducing factor is 0.1926. For the 21-point rule, the error reduction is 0.1432. In general, *longer* graduation rules have smaller variance reducing factors.

The systematic error $\mu_j - \sum l_k \mu_{j+k}$ cannot be characterized without knowing μ. But for cubic reproducing rules and sufficiently nice μ, the dominant term of the systematic error arises from the $O(\nabla^4 y_j)$ term in (1.4). This can be found explicitly, either by continuing the expansion (1.3), or graduating $y_j = j^4/24$ (Exercise 1.2). For the 15-point rule, $\hat{y}_j = y_j - 3.8625\nabla^4 y_j + O(\nabla^6 y_j)$. For the 21-point rule, $\hat{y}_j = y_j - 12.6\nabla^4 y_j + O(\nabla^6 y_j)$. In general, *shorter* graduation rules have smaller systematic error.

Clearly, choosing the length of a graduation rule, or *bandwidth*, involves a compromise between systematic error and random error. Largely, the choice can be guided by graphical techniques and knowledge of the problem at hand. For example, we expect mortality rates, such as those in Figure 1.1, to be a monotone increasing function of age. If the results of a graduation were not monotone, one would try a longer graduation rule. On the other hand if the graduation shows systematic error, with several successive points lie on one side of the fitted curve, this indicates that a shorter graduation rule is needed.

1.2   Local Polynomial Fitting

The summation formulae are motivated by their cubic reproduction property and the simple sequence of arithmetic operations required for their computation. But Henderson (1916) took a different approach. Define a sequence of non-negative weights $\{w_k\}$, and solve the system of equations

$$\sum w_k(a + bk + ck^2 + dk^3) = \sum w_k y_{j+k}$$
$$\sum kw_k(a + bk + ck^2 + dk^3) = \sum kw_k y_{j+k}$$
$$\sum k^2 w_k(a + bk + ck^2 + dk^3) = \sum k^2 w_k y_{j+k}$$
$$\sum k^3 w_k(a + bk + ck^2 + dk^3) = \sum k^3 w_k y_{j+k} \qquad (1.7)$$

for the unknown coefficients a, b, c, d. Thus, a cubic polynomial is fitted to the data, locally within a neighborhood of y_j. The graduated value $\hat{y}_j$ is then the coefficient a. Clearly this is cubic-reproducing, since if $y_{j+k} = a + bk + ck^2 + dk^3$ both sides of (1.7) are identical. Also note the local cubic method provides graduated values right up to the boundaries; this is more appealing than the extrapolation method we used with Spencer's formulae.

Henderson showed that the weight diagram $\{l_k\}$ for this procedure is simply w_k multiplied by a cubic polynomial. More importantly, he also showed a converse. If the weight diagram of a cubic-reproducing graduation formula has at most three sign changes, then it can be interpreted as a local cubic fit with an appropriate sequence of weights w_k. The route from $\{l_k\}$ to $\{w_k\}$ is quite explicit: Divide by a cubic polynomial whose roots match those of $\{l_k\}$. For Spencer's 15-point rule, the roots of the weight diagram (1.5) lie between 4 and 5, so dividing by $20 - k^2$ gives appropriate weights for a local cubic polynomial.

1.2.1   Optimal Weights

For a fixed constant $m \geq 1$, consider the weight diagram

$$l_k^0 = \frac{3}{(2m+1)(4m^2 - 4m - 3)}(3m^2 + 3m - 1 - 5k^2) \qquad (1.8)$$

for $|k| \leq m$, and 0 otherwise. It can be verified that $\{l_k^0\}$ satisfies the cubic reproduction property (1.6). Note that by Henderson's representation, $\{l_k^0\}$ is local cubic regression, with $w_k = 1$ for $|k| \leq m$. Now let $\{l_k\}$ be any other weight diagram supported on $[-m, m]$, also satisfying the constraints (1.6). Writing $l_k = l_k^0 + (l_k - l_k^0)$ yields

$$\sum_{k=-m}^{m} l_k^2 = \sum_{k=-m}^{m} (l_k^0)^2 + \sum_{k=-m}^{m} (l_k - l_k^0)^2 + 2 \sum_{k=-m}^{m} l_k^0(l_k - l_k^0). \qquad (1.9)$$

Note that $\{l_k^0\}$ is a quadratic (and cubic) polynomial; $l_k^0 = P(k)$. The final sum can be written as

$$\sum_{k=-m}^{m} l_k^0(l_k - l_k^0) = \sum_{k=-m}^{m} P(k)(l_k - l_k^0).$$

Using the cubic reproduction property of both $\{l_k\}$ and $\{l_k^0\}$,

$$\sum_{k=-m}^{m} P(k)l_k - \sum_{k=-m}^{m} P(k)l_k^0 = P(0) - P(0) = 0.$$

Substituting this in (1.9) yields

$$\sum_{k=-m}^{m} l_k^2 = \sum_{k=-m}^{m} (l_k^0)^2 + \sum_{k=-m}^{m} (l_k - l_k^0)^2$$
$$\geq \sum_{k=-m}^{m} (l_k^0)^2.$$

That is, $\{l_k^0\}$ minimizes the variance reducing factor among all cubic reproducing weight diagrams supported on $[-m, m]$. This optimality property was discussed by several authors, including Schiaparelli (1866), De Forest (1877) and Sheppard (1914a,b).

Despite minimizing the variance reducing factor, the weight diagram (1.8) can lead to rough graduations, since as j changes, observations rapidly switch into and out of the window $[j - m, j + m]$. This led several authors to derive graduation rules minimizing the variance of higher order differences of the graduated values, subject to polynomial reproduction. Borgan (1979) discusses some of the history of these results.

The first results of this type were in De Forest (1873), who minimized the variances of the fourth differences $\nabla^4 \hat{y}_j$, subject to the cubic reproduction property. Explicit solutions were given only for small values of m.

Henderson (1916) measured the amount of smoothing by variance of the third differences $\nabla^3 \hat{y}_j$, subject to cubic reproduction. Equivalently, one minimizes the sum of squares of third differences of the weight diagram, $\sum(\nabla^3 l_k)^2$. The solution, which became known as Henderson's ideal formula, was a local cubic smooth with weights

$$w_k = ((m+1)^2 - k^2)((m+2)^2 - k^2)((m+3)^2 - k^2); k = -m, \ldots, m.$$

For example, for $m = 7$, this produces the 15-point rule with weight diagram

$$\{l_k\}_{k=-7}^{7} = (-0.014, -0.024, -0.014, 0.024, 0.083, 0.146, 0.194, 0.212,$$
$$0.194, 0.146, 0.083, 0.024, -0.014, -0.024, -0.014).$$

Remark. The optimality results presented here have been rediscovered several times in modern literature, usually in asymptotic variants. Henderson's ideal formula is a finite sample variant of the $(0, 4, 3)$ kernel in Table 1 of Müller (1984); see Exercise 1.6.

1.3   Smoothing of Time Series

Smoothing methods have been widely used to estimate trends in economic time series. A starting point is the book Macaulay (1931), which was heavily influenced by the work of Henderson and other actuaries. Many books on time series analysis discuss smoothing methods, for example, chapter 3 of Anderson (1971) or chapter 3 of Kendall and Ord (1990).

Perhaps the most notable effort in time series occurred at the U. S. Bureau of the Census. Beginning in 1954, the bureau developed a series of computer programs for seasonal adjustment of time series. The X-11 method uses moving averages to model seasonal effects, long-term trends and trading day effects in either additive or multiplicative models. A full technical description of X-11 is Shiskin, Young and Musgrave (1967); the main features are also discussed in Wallis (1974), Kenny and Durbin (1982) and Kendall and Ord (1990).

The X-11 method provides the first computer implementation of smoothing methods. The algorithm alternately estimates trend and seasonal components using moving averages, in a manner similar to what is now known as the backfitting algorithm (Hastie and Tibshirani 1990).

X-11 also incorporates some other notable contributions. The first is robust smoothing. At each stage of the estimation procedure, X-11 identifies observations with large irregular (or residual) components, which may unduly influence the trend estimates. These observations are then shrunk toward the moving average.

Another contribution of X-11 is data-based bandwidth selection, based on a comparison of the smoothness of the trend and the amount of random fluctuation in the series. After seasonal adjustment of the series, Henderson's ideal formula with 13 terms ($m = 6$) is applied. The average absolute month-to-month changes are computed, for both the trend estimate and the irregular (residual) component. Let these averages be $\bar{C}$ and $\bar{I}$ respectively, so $\bar{I}/\bar{C}$ is a measure of the noise-to-signal ratio. If $\bar{I}/\bar{C} < 1$, this indicates the sequence has low noise, and the trend estimate is recomputed with 9 terms. If $\bar{I}/\bar{C} \geq 3.5$, the sequence has high noise, and the trend estimate is recomputed with 23 terms.

The time series literature also gave rise to a second smoothing problem. In spectral analysis, one expresses a time series as a sum of sine and cosine terms, and the spectral density (or periodogram) represents a decomposition of the sum of squares into terms represented at each frequency. It turns out that the sample spectral density provides an unbiased, but not consistent, estimate of the population spectral density. Consistency can be achieved by smoothing the sample spectral density. Various methods of local averaging were considered by Daniell (1946), Bartlett (1950), Grenander and Rosenblatt (1953), Blackman and Tukey (1958), Parzen (1961) and others. Local polynomial methods were applied to this problem by Daniels (1962).

1.4   Modern Local Regression

The importance of local regression and smoothing methods is demonstrated by the number of different fields in which the methods have been applied. Early contributions were made in fields as diverse as astronomy, actuarial science and economics. Modern areas of application include numerical analysis (Lancaster and Salkauskas 1986), sociology (Wu and Tuma 1990), economics (Cowden 1962; Shiskin, Young and Musgrave 1967; Kenny and Durbin 1982), chemometrics (Savitzky and Golay 1964, Wang, Isaksson and Kowalski 1994), computer graphics (McLain 1974) and machine learning (Atkeson, Moore and Schaal 1997).

Despite the long history, local regression methods received little attention in the statistics literature until the late 1970s. Independent work around that time includes the mathematical development of Stone (1977), Katkovnik (1979) and Stone (1980), and the LOWESS procedure of Cleveland (1979). The LOWESS procedure was widely adopted in statistical software as a standard for estimating smooth functions.

The local regression method has been developed largely as an extension of parametric regression methods, and is accompanied by an elegant *finite sample* theory of linear estimation that builds on theoretical results for parametric regression. The work was initialized in some of the papers mentioned above and in the early work of Henderson. The theory was significantly developed in the book by Katkovnik (1985), and by Cleveland and Devlin (1988). Linear estimation theory also heavily uses ideas developed in the spline smoothing literature (Wahba 1990), particularly in the area of goodness of fit statistics and model selection.

Among other features, the local regression method and linear estimation theory trivialize problems that have proven to be major stumbling blocks for more widely studied kernel methods. The kernel estimation literature contains extensive work on bias correction methods: finding modifications that *asymptotically* remove dependence of the bias on slope, curvature and so forth. Examples include boundary kernels (Müller 1984), double smoothing (Härdle, Hall and Marron 1992), reflection methods (Hall and Wehrly 1991) and higher order kernels (Gasser, Müller and Mammitzsch 1985). But local regression trivially provides a *finite sample* solution to these problems. Local linear regression reproduces straight lines, so the bias cannot depend on the first derivative of the mean function. Local quadratic regression reproduces quadratics, so the bias cannot depend on the second derivative. And so on. Hastie and Loader (1993) contains an extensive discussion of these issues.

An alternative theoretical treatment of local regression is to view the method as an extension of kernel methods and attempt to extend the theory of kernel methods. This treatment has become popular in recent years, for example in Wand and Jones (1995) and to some extent in Fan and Gijbels (1996). The approach has its uses: Small bandwidth asymptotic properties of local regression, such as rates of convergence and optimality theory, rely heavily on results for kernel methods. But for practical purposes, the kernel theory is of limited use, since it often provides poor approximations and requires restrictive conditions.

There are many other procedures for fitting curves to data and only a few can be mentioned here. Smoothing spline and penalized likelihood methods were introduced by Whitaker (1923) and Henderson (1924a). In modern literature there are several distinct smoothing approaches using splines; references include Wahba (1990), Friedman (1991), Dierckx (1993), Green and Silverman (1994), Eilers and Marx (1996) and Stone, Hansen, Kooperberg and Truong (1997).

Orthogonal series methods such as wavelets (Donoho and Johnstone 1994) transform the data to an orthonormal set of basis functions, and retain basis functions with sufficiently large coefficients. The methods are particularly suited to problems with sharp features, such as spikes and discontinuities.

For high dimensional problems, many approaches based on dimension reduction have been proposed: Projection pursuit (Friedman and Stuetzle 1981); regression trees (Breiman, Friedman, Olshen and Stone 1984), additive models (Breiman and Friedman 1985; Hastie and Tibshirani 1986) among others. Neural networks have become popular in recent years in computer science, engineering and other fields. Cheng and Titterington (1994) provide a statistical perspective and explore further the relation between neural networks and statistical curve fitting procedures.

1.5   Exercises

1.1 Consider the graduation rule

$$\hat{y}_j = \frac{S_{k,k}}{k^2} y_j.$$

That is, $\hat{y}_j$ is formed by two successive moving averages of length k.

a) Let $k = 4$. Show the graduation rule has the explicit form

$$\hat{y}_j = \frac{1}{16}(y_{j-3} + 2y_{j-2} + 3y_{j-1} + 4y_j + 3y_{j+1} + 2y_{j+2} + y_{j+3}).$$

Show this has the difference representation

$$\hat{y}_j = y_j + \frac{5}{4}\nabla^2 y_j + \frac{1}{2}\nabla^4 y_j + \frac{1}{16}\nabla^6 y_j.$$

b) Let $y_j = j^2/2$. For general k, show

$$\hat{y}_0 = \frac{k^2 - 1}{12}.$$

c) Let $y_j = (j^4 - j^2)/24$. For general k, show

$$\hat{y}_0 = \frac{(k^2 - 1)(k^2 - 4)}{360}.$$

d) Show

$$\hat{y}_j = (I + \frac{k^2 - 1}{12}\nabla^2 + \frac{(k^2 - 1)(k^2 - 4)}{360}\nabla^4 + O(\nabla^6))y_j.$$

Using the series expansion $\sqrt{1 + x} = 1 + x/2 - x^2/8 + O(x^3)$, establish Theorem 1.1 for general k.

The following results may be useful:

$$\sum_{j=0}^{k-1} j = \frac{k(k-1)}{2}$$

$$\sum_{j=0}^{k-1} j^2 = \frac{k(k-1)(2k-1)}{6}$$

$$\sum_{j=0}^{k-1} j^3 = \frac{k^2(k-1)^2}{4}$$

$$\sum_{j=0}^{k-1} j^4 = \frac{k(k-1)(2k-1)(3k^2 - 3k - 1)}{30}$$

$$\sum_{j=0}^{k-1} j^5 = \frac{k^2(k-1)^2(2k^2 - 2k - 1)}{12}$$

1.2   a) Show the weight diagram for any graduation rule can be found by applying the graduation rule to the unit vector

$$(\ldots \quad 0 \quad 0 \quad 1 \quad 0 \quad 0 \quad \ldots).$$

Compute the weight diagram for Spencer's 21-point rule, for Woolhouse's (1870) rule

$$\hat{y}_j = \frac{S_{5,5,5}}{125}(-3y_{j-1} + 7y_j - 3y_{j+1})$$

and for Higham's rule

$$\hat{y}_j = \frac{S_{5,5,5}}{125}(-y_{j-2} + y_{j-1} + y_j + y_{j+1} - y_{j+2}).$$

b) For a cubic reproducing graduation rule, show that the coefficient of ∇^4 in the difference expansion can be found as the graduated value of $y_j = j^4/24$ at $j = 0$.

1.3 Suppose a graduation rule has a weight diagram with all positive weights $l_j \geq 0$ and that it reproduces constants (i.e. $\sum l_j = 1$). Also assume $l_j \neq 0$ for some $j \neq 0$. Show that graduation rule cannot be cubic reproducing. That is, there exists a cubic (or lower degree) polynomial that will not be reproduced by the graduation rule.

1.4 Compute the error reduction factors and coefficients of ∇^4 for Henderson's formula with $m = 5, \ldots, 10$. Make a scatterplot of the two components. Also compute and add the corresponding points for Spencer's 15- and 21-point rules, Woolhouse's rule and Higham's rule.

Remark. This exercise shows the bias-variance trade-off: As the length of the graduation rule increases, the variance decreases but the coefficient of $\nabla^4 y_j$ increases (in absolute value).

1.5 For each year in the age range 20 to 45, 1000 customers each wish to buy a \$10000 life insurance policy. Two competing companies set premiums as follows: First, estimate the mortality rate for each age, then set the premium to cover the expected payout, plus a 10% profit. For example, if the company estimates 40 year olds to have a mortality rate of 0.01, the expected (per customer) payout is $0.01 \times \$10000 = \$100$, so the premium is \$110. Both companies use Spencer's mortality data to estimate mortality rates. The Gauss Life Company uses a least squares fit to the data, while Spencer Underwriting applies Spencer's 15-point rule.

a) Compute for each age group the premiums charged by each company.

b) Suppose perfect customer behavior, so, for example, all the 40 year old customers choose the company offering the lowest premium to 40 year olds. Also suppose Spencer's 21-point rule provides the true mortality rates. Under these assumptions, compute the expected profit (or loss) for each of the two companies.

1.6 For large m, show the weights for Henderson's ideal formula are approximately $m^6 W(k/m)$ where $W(v) = (1 - x^2)_+^3$. Thus, conclude that the weight diagram is approximately $315/512 \times W(k/m)(3 - 11(k/m)^2)$. Compare with the $(0, 4, 3)$ kernel in Table 1 of Müller (1984).

2
Local Regression Methods

This chapter introduces the basic ideas of local regression and develops important methodology and theory. Section 2.1 introduces the local regression method. Sections 2.2 and 2.3 discuss, in a mostly nontechnical manner, statistical modeling issues. Section 2.2 introduces the bias-variance trade-off and the effect of changing smoothing parameters. Section 2.3 discusses diagnostic techniques, such as residual plots and confidence intervals. Section 2.4 introduces more formal criteria for model comparison and selection, such as cross validation.

The final two sections are more technical. Section 2.5 introduces the theory of linear estimation. This provides characterizations of the local regression estimate and studies some properties of the bias and variance. Section 2.6 introduces asymptotic theory for local regression.

2.1 The Local Regression Estimate

Local regression is used to model a relation between a predictor variable (or variables) x and response variable Y, which is related to the predictor variables. Suppose a dataset consists of n pairs of observations, $(x_1, Y_1), (x_2, Y_2), \ldots, (x_n, Y_n)$. We assume a model of the form

$$Y_i = \mu(x_i) + \epsilon_i \tag{2.1}$$

where $\mu(x)$ is an unknown function and ϵ_i is an error term, representing random errors in the observations or variability from sources not included in the x_i.

The errors ϵ_i are assumed to be independent and identically distributed with mean 0; $E(\epsilon_i) = 0$, and have finite variance; $E(\epsilon_i^2) = \sigma^2 < \infty$. Globally, no strong assumptions are made about μ. Locally around a point x, we assume that μ can be well approximated by a member of a simple class of parametric functions. For example, Taylor's theorem says that any differentiable function can be approximated locally by a straight line, and a twice differentiable function can be approximated by a quadratic polynomial.

For a fitting point x, define a bandwidth $h(x)$ and a smoothing window $(x - h(x), x + h(x))$. To estimate $\mu(x)$, only observations within this window are used. The observations weighted according to a formula

$$w_i(x) = W\left(\frac{x_i - x}{h(x)}\right) \tag{2.2}$$

where $W(u)$ is a weight function that assigns largest weights to observations close to x. For many of our examples, we use the tricube weight function

$$W(u) = (1 - |u|^3)^3. \tag{2.3}$$

Within the smoothing window, $\mu(u)$ is approximated by a polynomial. For example, a local quadratic approximation is

$$\mu(u) \approx a_0 + a_1(u - x) + \frac{1}{2}a_2(u - x)^2 \tag{2.4}$$

whenever $|u - x| < h(x)$. A compact vector notation for polynomials is

$$a_0 + a_1(u - x) + \frac{1}{2}a_2(u - x)^2 = \langle a, A(u - x)\rangle$$

where a is a vector of the coefficients and $A(\cdot)$ is a vector of the fitting functions. For local quadratic fitting,

$$a = \begin{pmatrix} a_0 \\ a_1 \\ a_2 \end{pmatrix} \qquad A(v) = \begin{pmatrix} 1 \\ v \\ \frac{v^2}{2} \end{pmatrix}.$$

The coefficient vector a can be estimated by minimizing the locally weighted sum of squares:

$$\sum_{i=1}^{n} w_i(x)(Y_i - \langle a, A(x_i - x)\rangle)^2. \tag{2.5}$$

The local regression estimate of $\mu(x)$ is the first component of $\hat{a}$.

Definition 2.1 The **local regression estimate** is

$$\hat{\mu}(x) = \langle \hat{a}, A(0)\rangle = \hat{a}_0, \tag{2.6}$$

obtained by setting $u = x$ in (2.4).

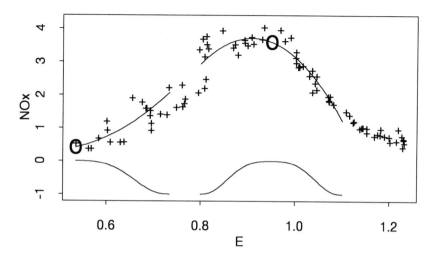

FIGURE 2.1. Local regression: Smoothing windows (bottom); local least squares fits (solid curves) and estimates $\hat{\mu}(x)$ (big circles).

The local regression procedure is illustrated in Figure 2.1. The ethanol dataset, measuring exhaust emissions of a single cylinder engine, is originally from Brinkman (1981) and has been studied extensively by Cleveland (1993) and others. The response variable, NOx, measures the concentration of certain pollutants in the emissions, and the predictor variable, E, is the equivalence ratio, measuring the richness of the air and fuel mix in the engine. Figure 2.1 illustrates the fitting procedure at the points $E = 0.535$ and $E = 0.95$. The observations are weighted according to the two weight functions shown at the bottom of Figure 2.1. The local quadratic polynomials are then fitted within the smoothing windows. From each quadratic, only the central point, indicated by the large circles in Figure 2.1, is retained. As the smoothing window slides along the data, the fitted curve is generated. Figure 2.2 displays the resulting fit.

The preceding demonstration has used local quadratic polynomials. It is instructive to consider lower order fits.

Example 2.1. (Local Constant Regression) For local constant polynomials, there is just one local coefficient a_0, and the local residual sum of squares (2.5) is

$$\sum_{i=1}^{n} w_i(x)(Y_i - a_0)^2.$$

The minimizer is easily shown to be

$$\hat{\mu}(x) = \hat{a}_0 = \frac{\sum_{i=1}^{n} w_i(x)Y_i}{\sum_{i=1}^{n} w_i(x)}. \tag{2.7}$$

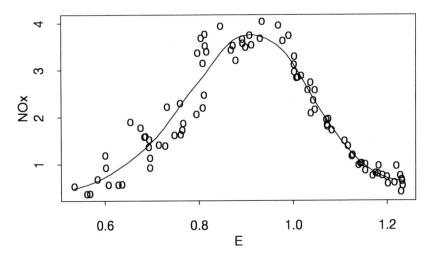

FIGURE 2.2. Local regression fit of the ethanol data.

This is the kernel estimate of Nadaraya (1964) and Watson (1964). It is simply a weighted average of observations in the smoothing window. A local constant approximation can often only be used with small smoothing windows, and noisy estimates result. The estimate is particularly susceptible to boundary bias. In Figure 2.1, if a local constant fit was used at $E = 0.535$, it would clearly lie well above the data.

Example 2.2. (Local Linear Regression) The local linear estimate, with $A(v) = (1 \quad v)^T$, has the closed form

$$\hat{\mu}(x) = \frac{\sum_{i=1}^{n} w_i(x)Y_i}{\sum_{i=1}^{n} w_i(x)} + (x - \bar{x}_w)\frac{\sum_{i=1}^{n} w_i(x)(x_i - \bar{x}_w)Y_i}{\sum_{i=1}^{n} w_i(x)(x_i - \bar{x}_w)^2} \qquad (2.8)$$

where $\bar{x}_w = \sum_{i=1}^{n} w_i(x)x_i / \sum_{i=1}^{n} w_i(x)$. See exercise 2.1. That is, the local linear estimate is the local constant estimate, plus a correction for local slope of the data and skewness of the x_i. This correction reduces the boundary bias problem of local constant estimates. When the fitting point x is *not* near a boundary, one usually has $x \approx \bar{x}_w$, and there is little difference between local constant and local linear fitting. A local linear estimate exhibits bias if the mean function has high curvature.

2.1.1 Interpreting the Local Regression Estimate

In studies of linear regression, one often focuses on the regression coefficients. One assumes the model being fitted is correct and asks questions such as how well the estimated coefficients estimate the true coefficients. For example, one might compute variances and confidence intervals for the regression coefficients, test significance of the coefficients or use model selection criteria, such as stepwise selection, to decide what coefficients to include in the model. The fitted curve itself often receives relatively little attention.

In local regression, we have to change our focus. Instead of concentrating on the coefficients, we focus on the fitted curve. A basic question that can be asked is "how well does $\hat{\mu}(x)$ estimate the true mean $\mu(x)$?". When variance estimates and confidence intervals are computed, they will be computed for the curve estimate $\hat{\mu}(x)$. Model selection criteria can still be used to select variables for the local model. But they also have a second use, addressing whether an estimate $\hat{\mu}(x)$ is satisfactory or whether alternative local regression estimates, for example, with different bandwidths, produce better results.

2.1.2 Multivariate Local Regression

Formally, extending the definition of local regression to multiple predictors is straightforward; we require a multivariate weight function and multivariate local polynomials. This was considered by McLain (1974) and Stone (1982). Statistical methodology and visualization for multivariate fitting was developed by Cleveland and Devlin (1988) and the associated LOESS method.

With two predictor variables, the local regression model becomes

$$Y_i = \mu(x_{i,1}, x_{i,2}) + \epsilon_i,$$

where $\mu(\cdot, \cdot)$ is unknown. Again, a suitably smooth function μ can be approximated in a neighborhood of a point $x = (x_{.,1}, x_{.,2})$ by a local polynomial; for example, a local quadratic approximation is

$$\mu(u_1, u_2) \approx a_0 + a_1(u_1 - x_{.,1}) + a_2(u_2 - x_{.,2}) + \frac{a_3}{2}(u_1 - x_{.,1})^2$$
$$+ a_4(u_1 - x_{.,1})(u_2 - x_{.,2}) + \frac{a_5}{2}(u_2 - x_{.,2})^2.$$

This can again be written in the compact form

$$\mu(u_1, u_2) \approx \langle a, A(u - x) \rangle,$$

where $A(\cdot)$ is the vector of local polynomial basis functions:

$$A\begin{pmatrix} v_1 \\ v_2 \end{pmatrix} = \begin{pmatrix} 1 \\ v_1 \\ v_2 \\ \frac{1}{2}v_1^2 \\ v_1 v_2 \\ \frac{1}{2}v_2^2 \end{pmatrix}. \tag{2.9}$$

Weights are defined on the multivariate space, so observations close to a fitting point x receive the largest weight. First, define the length of a vector v in $\mathcal{R}^d$ by

$$\|v\|^2 = \sum_{j=1}^{d} \left(\frac{v_j}{s_j}\right)^2, \qquad (2.10)$$

where $s_j > 0$ is a scale parameter for the jth dimension. A spherically symmetric weight function gives an observation x_i the weight

$$w_i(x) = W\left(\frac{\|x_i - x\|}{h}\right). \qquad (2.11)$$

As in the univariate case, the local coefficients are estimated by solving the weighted least squares problem (2.5). Following Definition 2.1, $\hat{\mu}(x)$ is the first component of $\hat{a}$.

2.2   The Components of Local Regression

Much work remains to be done to make local regression useful in practice. There are several components of the local fit that must be specified: the bandwidth, the degree of local polynomial, the weight function and the fitting criterion.

2.2.1   Bandwidth

The bandwidth $h(x)$ has a critical effect on the local regression fit. If $h(x)$ is too small, insufficient data fall within the smoothing window, and a noisy fit, or large variance, will result. On the other hand, if $h(x)$ is too large, the local polynomial may not fit the data well within the smoothing window, and important features of the mean function $\mu(x)$ may be distorted or lost completely. That is, the fit will have large bias. The bandwidth must be chosen to compromise this bias-variance trade-off.

Ideally, one might like to choose a separate bandwidth for each fitting point, taking into account features such as the local density of data and the amount of structure in the mean function. In practice, doing this in a sensible manner is difficult. Usually, one restricts attention to bandwidth functions with a small number of parameters to be selected.

The simplest specification is a **constant bandwidth**, $h(x) = h$ for all x. This is satisfactory in some simple examples, but when the independent variables x_i have a nonuniform distribution, this can obviously lead to problems with empty neighborhoods. This is particularly severe in boundary or tail regions or in more than one dimension.

Data sparsity problems can be reduced by ensuring neighborhoods contain sufficient data. A **nearest neighbor bandwidth** chooses $h(x)$ so that

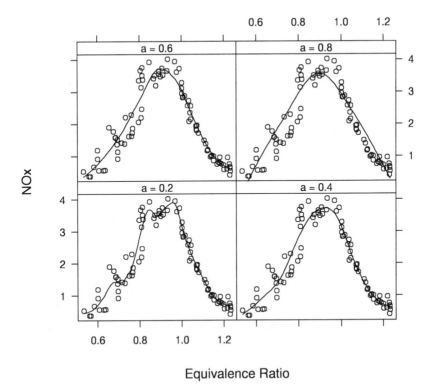

Equivalence Ratio

FIGURE 2.3. Local fitting at different bandwidths. Four different nearest neighbor fractions: $\alpha = 0.8, 0.6, 0.4$ and 0.2 are used.

the local neighborhood always contains a specified number of points. For a smoothing parameter α between 0 and 1, the nearest neighbor bandwidth $h(x)$ is computed as follows:

1. Compute the distances $d(x, x_i) = |x - x_i|$ between the fitting point x and the data points x_i.

2. Choose $h(x)$ to be the kth smallest distance, where $k = \lfloor n\alpha \rfloor$.

Example 2.3. Figure 2.3 shows local quadratic fits for the ethanol dataset using four different values of α. Clearly, the fit produced by the smallest fraction, $\alpha = 0.2$, produces a much noisier fit than the largest bandwidth, $\alpha = 0.8$. In fact, $\alpha = 0.8$ has oversmoothed, since it doesn't track the data well. For $1.0 < E < 1.2$, there is a sequence of 17 successive data points lying below the fitted curve. The leveling off at the right boundary is not captured. The peak for $0.9 < E < 1.0$ appears to be trimmed.

The fit with $\alpha = 0.2$ shows features - bimodality of the peak and a leveling off around $E = 0.7$ that don't show up at larger bandwidths. Are these additional features real, or are they artifacts of random noise in the data? Our *apriori guess* might be that these are random noise; we hope that nature isn't too nasty. But *proving* this from the data is impossible. There are small clumps of observations that support both of the additional features in the plot with $\alpha = 0.2$, but probably not enough to declare statistical significance.

This example is discussed in more detail later. For now, we note the one-sided nature of bandwidth selection. While large smoothing parameters may easily be rejected as oversmoothed, it is much more difficult to conclude *from the data alone* that a small bandwidth is undersmoothed.

2.2.2 Local Polynomial Degree

Like the bandwidth, the degree of the local polynomial used in (2.5) affects the bias-variance trade-off. A high polynomial degree can always provide a better approximation to the underlying mean $\mu(u)$ than a low polynomial degree. Thus, fitting a high degree polynomial will usually lead to an estimate $\hat{\mu}(x)$ with less bias. But high order polynomials have large numbers of coefficients to estimate, and the result is variability in the estimate. To some extent, the effects of the polynomial degree and bandwidth are confounded. For example, if a local quadratic estimate and local linear estimate are computed using the same bandwidth, the local quadratic estimate will be more variable. But the variance increase can be compensated by increasing the bandwidth.

It often suffices to choose a low degree polynomial and concentrate on choosing the bandwidth to obtain a satisfactory fit. The most common choices are local linear and local quadratic. As noted in Example 2.1, a local constant fit is susceptible to bias and is rarely adequate. A local linear estimate usually produces better fits, especially at boundaries. A local quadratic estimate reduces bias further, but increased variance can be a problem, especially at boundaries. Fitting local cubic and higher orders rarely produces much benefit.

Example 2.4. Figure 2.4 displays local constant, local linear, local quadratic and local cubic fits for the ethanol dataset. Nearest neighbor bandwidths are used, with $\alpha = 0.25, 0.3, 0.49$ and 0.59 for the four degrees. These smoothing parameters are chosen so that each fit has about seven degrees of freedom; a concept defined in section 2.3.2. Roughly, two fits with the same degrees of freedom have the same variance $\text{var}(\hat{\mu}(x))$.

The local constant fit in Figure 2.4 is quite noisy, and also shows boundary bias: The fit doesn't track the data well at the left boundary. The local linear fit reduces both the boundary bias and the noise. A closer examination suggests the local constant and linear fit have trimmed the peak: For $0.8 < E < 1.0$, nearly all the data points are *above* the fitted curve. The

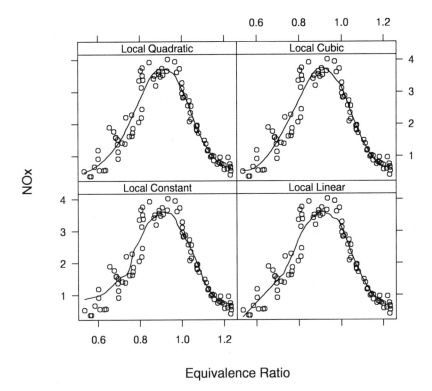

FIGURE 2.4. Ethanol data: Effect of changing the polynomial degree.

local quadratic and local cubic fits in Figure 2.4 produce better results: The fits show less noise and track the data better.

2.2.3   The Weight Function

The weight function $W(u)$ has much less effect on the bias-variance trade-off, but it influences the visual quality of the fitted regression curve. The simplest weight function is the rectangular:

$$W(u) = I_{[-1,1]}(u).$$

This weight function is rarely used, since it leads to discontinuous weights $w_i(x)$ and a discontinuous fitted curve. Usually, $W(u)$ is chosen to be continuous, symmetric, peaked at 0 and supported on $[-1, 1]$. A common choice is the tricube weight function (2.3).

Other types of weight function can also be useful. Friedman and Stuetzle (1982) use smoothing windows covering the same number of data points both before and after the fitting point. For nonuniform designs this is asymmetric, but it can improve variance properties. McLain (1974) and Lancaster and Salkaus kas (1981) use weight functions with singularities at $u = 0$. This leads to a fitted smooth curve that interpolates the data. In Section 6.3, one-sided weight functions are used to model discontinuous curves.

2.2.4   The Fitting Criterion

The local regression estimate, as defined by (2.5) and (2.6), is a local least squares estimate. This is convenient, since the estimate is easy to compute and much of the methodology available for least squares methods can be extended fairly directly to local regression. But it also inherits the bad properties of least squares estimates, such as sensitivity to outliers.

Any other fitting criterion can be used in place of least squares. The local likelihood method uses likelihoods instead of least squares; this forms a major topic later in this book. Local robust regression methods are discussed in Section 6.4.

2.3   Diagnostics and Goodness of Fit

In local regression studies, one is faced with several model selection issues: Variable selection, choice of local polynomial degree and smoothing parameters. An ideal aim may be fully automated methods: We plug data into a program, and it automatically returns the best fit. But this goal is unattainable, since the best fit depends not only on the data, but on the questions of interest.

What statisticians (and statistical software) can provide is tools to help guide the choice of smoothing parameters. In this section we introduce some graphical aids to help the decision: residual plots, degrees of freedom and confidence intervals. Some more formal tools are introduced in Section 2.4. These tools are designed to help decide which features of a dataset are real and which are random. They cannot provide a definitive answer as to the best bandwidth for a (dataset,question) pair.

The ideas for local regression are similar to those used in parametric models. Other books on regression analysis cover these topics in greater detail than we do here; see, for example, chapter 3 of Draper and Smith (1981) or chapters 4, 5 and 6 of Myers (1990). Cleveland (1993) is a particularly good reference for graphical diagnostics.

It is important to remember that no one diagnostic technique will explain the whole story of a dataset. Rather, using a combination of diagnostic tools and looking at these in conjunction with both the fitted curves and the original data provide insight into the data. What features are real;

have these been adequately modeled; are underlying assumptions, such as homogeneity of variance, satisfied?

2.3.1 Residuals

The most important diagnostic component is the residuals. For local regression, the residuals are defined as the difference between observed and fitted values:

$$\hat{\epsilon}_i = Y_i - \hat{\mu}(x_i).$$

One can use the residuals to construct formal tests of goodness of fit or to modify the local regression estimate for nonhomogeneous variance. These topics will be explored more in Chapter 9. For practical purposes, most insight is often gained simply by plotting the residuals in various manners. Depending on the situation, plots that can be useful include:

1. Residuals vs. predictor variables, for detecting lack of fit, such as a trimmed peak.

2. Absolute residuals vs. the predictors, to detect dependence of residual variance on the predictor variables. One can also plot absolute residuals vs. fitted values, to detect dependence of the residual variance on the mean response.

3. Q-Q plots (Wilk and Gnanadesikan 1968), to detect departure from normality, such as skewness or heavy tails, in the residual distribution. If non-normality is found, fitting criteria other than least squares may produce better results. See Section 6.4.

4. Serial plots of $\hat{\epsilon}_i$ vs. $\hat{\epsilon}_{i-1}$, to detect correlation between residuals.

5. Sequential plot of residuals, in the order the data were collected. In an industrial experiment, this may detect a gradual shift in experimental conditions over time.

Often, it is helpful to smooth residual plots: This can both draw attention to any features shown in the plot, as well as avoiding any visual pitfalls. Exercise 2.6 provides some examples where the wrong plot, or a poorly constructed plot, can provide misleading information.

Example 2.5. Figure 2.5 displays smoothed residual plots for the four fits in Figure 2.3. The residual plots are much better at displaying bias, or oversmoothing, of the fit. For example, the bias problems when $\alpha = 0.8$ are much more clearly displayed from the residual plots in Figure 2.5 than from the fits in Figure 2.3. Of course, as the smoothing parameter α is reduced, the residuals generally get smaller, and show less structure.

The smooths of the residuals in Figure 2.5 are constructed with $\alpha_r = 0.2$ (this should be distinguished from the α used to smooth the original data).

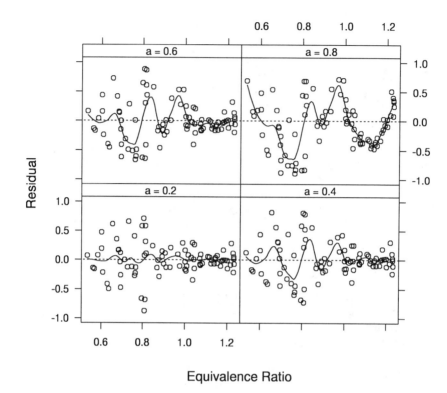

FIGURE 2.5. Residual plots for the ethanol dataset.

But α_r itself is not important. What is important is that the smooths help search for clusters of residuals that may indicate lack of fit. At $\alpha = 0.8$, the lack of fit is clear. At $\alpha = 0.6$ and $\alpha = 0.4$, the peaks in the smooth are generally supported by clumps of residuals, although generally not enough to indicate lack of fit.

This example shows that it is important not to look at the residual plots alone, but to use them in conjunction with plots of the fit. The object is to determine whether large residuals correspond to features in the data that have been inadequately modeled. The purpose of the plots can be related to the bias-variance trade-off:

- Plots of the fit help us detect noise in the fit.

- Residual plots help us detect bias.

It is important to note that the purpose of adding a smooth to a residual plot is not to provide a good estimate of the mean. Rather, it is to enhance our view of the residuals; by reducing the noise, our attention may be more readily drawn to features that have been missed or not properly modeled by the smooth.

2.3.2 Influence, Variance and Degrees of Freedom

How can we characterize the amount of smoothing being performed? The bandwidth provides one characterization. But this is not ideal, since it takes no account of the other choices that go into the smooth, such as the degree of local polynomial and the weight function. Moreover, the bandwidth doesn't enable meaningful comparison with other smoothing methods or with parametric models.

What we need is unitless characterizations which allow comparison between methods. We discuss two types of characterization:

- Pointwise criteria, characterizing the amount of smoothing at a single point. These include the variance reducing factor and influence function.

- Global criteria, characterizing the overall amount of smoothing. This is the fitted degrees of freedom.

Before proceeding with definitions, the importance of the ideas presented here, both in theory and practice, must be emphasized. Throughout this book these concepts (and generalizations) will appear repeatedly. We already saw the variance reducing factor used in optimality results in Chapter 1; this will also appear in inference and confidence interval construction. The influence function and degrees of freedom will appear repeatedly in model selection criteria. The importance of these concepts has of course been emphasized elsewhere, both in local regression literature and elsewhere in the smoothing literature. See Craven and Wahba (1979), Cleveland and Devlin (1988), Buja, Hastie and Tibshirani (1989), Wahba (1990), Hastie and Tibshirani (1990) and Cleveland and Loader (1996). Ye (1998) contains a nice discussion of the importance, motivation, generalizations and applications of degrees of freedom.

Because the local regression estimate solves a least squares problem, $\hat{\mu}(x)$ is a **linear estimate**. That is, for each x there exists a **weight diagram** vector $l(x) = \{l_i(x)\}_{i=1}^n$ such that

$$\hat{\mu}(x) = \sum_{i=1}^{n} l_i(x) Y_i. \tag{2.12}$$

For local constant regression, (2.7) gives the explicit formula

$$l_i(x) = \frac{w_i(x)}{\sum_{j=1}^{n} w_j(x)}.$$

For more general local regression, the weight diagram is derived in Section 2.5. The weight diagram leads to compact forms for the mean and variance of the local regression estimate:

$$E(\hat{\mu}(x)) = \sum_{i=1}^{n} l_i(x)\mu(x_i) = \langle l(x), \mu \rangle$$

$$\text{var}(\hat{\mu}(x)) = \sigma^2 \sum_{i=1}^{n} l_i(x)^2 = \sigma^2 \|l(x)\|^2. \qquad (2.13)$$

The variance assumes the observations Y_i are independent and have constant variance σ^2. The **variance reducing factor** $\|l(x)\|^2$ measures the reduction in variance due to the local regression. Usually, this decreases as the bandwidth increases. Under mild conditions, one can show (see Theorem 2.3):

$$\frac{1}{n} \le \|l(x_i)\|^2 \le l_i(x_i) \le 1. \qquad (2.14)$$

The extreme cases $1/n$ and 1 correspond, respectively, to $\hat{\mu}(x)$ being the sample average and interpolating the data.

The **hat matrix** is the $n \times n$ matrix $\mathbf{L}$ with rows $l(x_i)^T$, which maps the data to the fitted values:

$$\begin{pmatrix} \hat{\mu}(x_1) \\ \vdots \\ \hat{\mu}(x_n) \end{pmatrix} = \mathbf{L}Y. \qquad (2.15)$$

The **influence** or **leverage** values are the diagonal elements $l_i(x_i)$ of the hat matrix. We denote these by $\text{infl}(x_i)$; these measure the sensitivity of the fitted curve $\hat{\mu}(x_i)$ to the individual data points.

The **degrees of freedom** of a local fit provide a generalization of the number of parameters of a parametric model. In fact, there are several possible definitions, but two of the most useful are

$$\nu_1 = \sum_{i=1}^{n} \text{infl}(x_i) = \text{tr}(\mathbf{L})$$

$$\nu_2 = \sum_{i=1}^{n} \|l(x_i)\|^2 = \text{tr}(\mathbf{L}^T\mathbf{L}). \qquad (2.16)$$

The usefulness of the degrees of freedom is in providing a measure of the amount of smoothing that is comparable between different estimates applied to the same dataset. The concept was used in Example 2.4 to compare local polynomial fits of differing degrees, where we required smoothing parameters to perform the same amount of smoothing in each of the four cases.

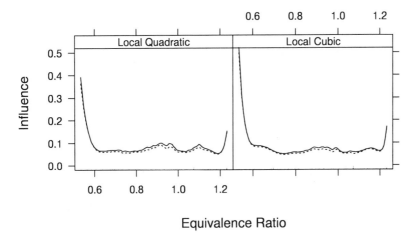

Equivalence Ratio

FIGURE 2.6. Influence functions (solid) and variance functions (dashed) for local quadratic and local cubic fits to the ethanol dataset.

For a parametric regression model, the hat matrix $\mathbf{L}$ is symmetric and idempotent, and the definitions coincide and usually equal the number of parameters. For local regression models, the two definitions are usually not equal; following (2.14), $1 \leq \nu_2 \leq \nu_1 \leq n$. Both of these definitions arise naturally later.

Example 2.6. Figure 2.6 shows the influence and variance functions for the local quadratic and local cubic fits from Figure 2.4. Largely, the influence values are slightly less than 0.1, indicating that Y_i constitutes about 10% of the fitted value $\hat{\mu}(x_i)$. The variance function is slightly less than the influence. The degrees of freedom are $\nu_1 = 7.16$ and $\nu_2 = 6.60$ for the local quadratic fit, and $\nu_1 = 6.97$ and $\nu_2 = 6.53$ for the local cubic.

But the main feature is the boundary effect, particularly at the left, where the influence function shows a huge increase. This reflects the difficulty of fitting a polynomial at boundary regions. Note also that the effect is more pronounced for the local cubic fit: This shows that boundaries are a main concern when choosing the degree of the local fit.

2.3.3   Confidence Intervals

If $\hat{\mu}(x)$ is an unbiased estimate of $\mu(x)$, an approximate confidence interval for the true mean is

$$I(x) = (\hat{\mu}(x) - c\hat{\sigma}\|l(x)\|, \hat{\mu}(x) + c\hat{\sigma}\|l(x)\|),$$

where c is the appropriate quantile of the standard normal distribution ($c = 1.96$ for 95% confidence) and $\hat{\sigma}$ is an estimate of the residual standard deviation.

Prediction intervals provide interval estimates for a new observation Y_{new} at a point x_{new}. Assuming the new observation is independent of the estimation data, one has

$$\text{var}(Y_{\text{new}} - \hat{\mu}(x_{\text{new}})) = \sigma^2(1 + \|l(x_{\text{new}})\|^2).$$

Thus, a prediction interval has limits

$$\hat{\mu}(x_{\text{new}}) \pm c\hat{\sigma}(1 + \|l(x)\|^2)^{1/2}. \tag{2.17}$$

Note that prediction intervals assume normality: If Y_{new} is not normally distributed, the prediction interval will not be correct, even asymptotically.

In analogy with parametric regression, the variance σ^2 can be estimated using the normalized residual sum of squares:

$$\hat{\sigma}^2 = \frac{1}{n - 2\nu_1 + \nu_2} \sum_{i=1}^{n} (Y_i - \hat{\mu}(x_i))^2, \tag{2.18}$$

where ν_1 and ν_2 are defined by (2.16). The residual degrees of freedom, $n - 2\nu_1 + \nu_2$, are defined so that $\hat{\sigma}^2$ is unbiased. See Section 2.5.1.

The assumption that $\hat{\mu}(x)$ is unbiased is rarely exactly true, so variance estimates and confidence intervals are usually computed at small bandwidths where bias is small. Confidence intervals, bias correction and variance estimation are discussed in more detail in Chapter 9.

2.4 Model Comparison and Selection

2.4.1 Prediction and Cross Validation

How good is a local regression estimate? To formalize this question, we need to define criteria with which to assess the performance of the fit. One possible criterion is the prediction mean squared error for future observations:

$$\text{PMSE}(\hat{\mu}) = \text{E}(Y_{\text{new}} - \hat{\mu}(x_{\text{new}}))^2. \tag{2.19}$$

Clearly, $\text{PMSE}(\hat{\mu})$ depends on assumptions made about x_{new}. For now, assume that the design points $x_1, \ldots, x_n$ are an independent sample from a density $f(x)$, and the new point x_{new} is sampled from the same density. The cross validation method provides an estimate of PMSE.

Definition 2.2 The **cross validation** estimate of the PMSE of an estimate $\hat{\mu}$ is

$$\text{CV}(\hat{\mu}) = \frac{1}{n} \sum_{i=1}^{n} (Y_i - \hat{\mu}_{-i}(x_i))^2 \tag{2.20}$$

where $\hat{\mu}_{-i}(x_i)$ denotes the leave-x_i-out estimate of $\mu(x_i)$. That is, each x_i is deleted from the dataset in turn, and the local regression estimate computed from the remaining $n - 1$ data points.

The leave-one-out cross validation criteria was introduced for parametric regression models by Allen (1974) as the PRESS (prediction error sum of squares) procedure. Wahba and Wold (1975) applied the method to smoothing splines. Model validation based on splitting datasets into *estimation data* and *prediction data* has a long history, discussed by Stone (1974) and Snee (1977) among others.

The generalized cross validation criterion was first proposed in the context of smoothing splines by Craven and Wahba (1979). This provides an approximation to cross validation and is easier to compute. The motivation for the definition will appear in Section 2.5.

Definition 2.3 The **generalized cross validation score** for a local estimate $\hat{\mu}$ is

$$\mathrm{GCV}(\hat{\mu}) = n\frac{\sum_{i=1}^{n}(Y_i - \hat{\mu}(x_i))^2}{(n - \nu_1)^2}, \qquad (2.21)$$

where ν_1 is the fitted degrees of freedom defined by (2.16).

2.4.2 Estimation Error and CP

The cross validation methods are motivated by prediction error: How well does $\hat{\mu}(x)$ predict new observations? Alternatively, one can consider estimation error: How well does $\hat{\mu}(x)$ estimate the true mean $\mu(x)$? One possible loss criterion is the sum of the squared error over the design points;

$$L(\hat{\mu}, \mu) = \sum_{i=1}^{n}(\hat{\mu}(x_i) - \mu(x_i))^2. \qquad (2.22)$$

The CP criterion, introduced by Mallows (1973) for parametric regression, provides an unbiased estimate of $L(\hat{\mu}, \mu)$ in the sense that $E(\mathrm{CP}(\hat{\mu})) = E(L(\hat{\mu}, \mu))$. The CP statistic was extended to local constant fitting by Rice (1984) and to local regression by Cleveland and Devlin (1988).

Definition 2.4 The **CP estimate of risk** for a local regression estimate $\hat{\mu}(x)$ is

$$\mathrm{CP}(\hat{\mu}) = \frac{1}{\sigma^2}\sum_{i=1}^{n}(Y_i - \hat{\mu}(x_i))^2 - n + 2\nu_1.$$

Implementation of the CP method requires an estimate of σ^2. The usual use of CP is to compare several different fits (for example, local regression with different bandwidths or different polynomial degrees). One should use

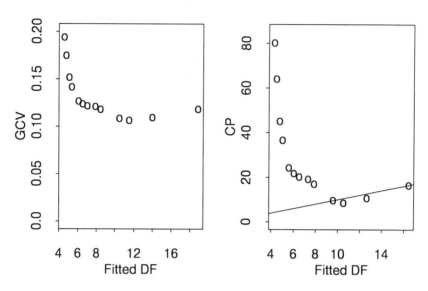

FIGURE 2.7. Generalized cross validation plot (left) and CP plot (right) for the ethanol dataset.

the *same* estimate $\hat{\sigma}^2$ for *all* fits being considered. The recommendation of Cleveland and Devlin is to compute the estimate (2.18) from a fit at the smallest bandwidth under consideration, at which one should be willing to assume that bias is negligible.

2.4.3   Cross Validation Plots

Frequently, the use of cross validation and CP is automated: A computer program computes $\text{CV}(\hat{\mu})$ or $\text{CP}(\hat{\mu})$ for several different fits and selects the fit with the lowest score. But, as argued strongly by Cleveland and Devlin (1988), this discards much of the information about the bias-variance trade-off that the statistics provide. Cleveland and Devlin introduce the CP (or M) plot as a graphical tool for displaying these statistics.

Example 2.7.   The GCV and CP statistics are computed for local quadratic fits to the ethanol dataset and a range of smoothing parameters; $0.2 \le \alpha \le 0.8$. The results are shown in Figure 2.7 as a **cross validation plot** (left) and **CP plot** (right). These plots use the fitted degrees of freedom $\text{tr}(\mathbf{L}^T\mathbf{L})$ as the horizontal axis and the GCV and CP statistics as the vertical axis. The smoothing parameter is $\alpha = 0.8$ on the left, decreasing in steps of 0.05 to $\alpha = 0.2$ on the right.

Both plots show a similar profile. The first four points, with fewer than five fitted degrees of freedom (or $\alpha > 0.65$), produce large GCV and CP scores, indicating these fits are inadequate. For larger degrees of freedom, the plots (especially GCV) are flat, indicating there is little to choose between the fits. As α is decreased from 0.6 to 0.2, the fitted degrees of freedom increases from 5.6 to 16.4, and the GCV score ranges from 0.107 to 0.127.

An important point in the construction of Figure 2.7 is the use of the fitted degrees of freedom, rather than the smoothing parameter, as the horizontal axis. This aids interpretation: Four degrees of freedom represents a smooth model with very little flexibility, while 16 degrees of freedom represents a noisy model showing many features. It also aids comparability. For example, CP scores could be computed for other polynomial degrees or for other smoothing methods and added to the plot.

The cross validation and CP plots must be emphasized as *a graphical aid in choosing smoothing parameters*. Flat plots, such as Figure 2.7, occur frequently, and any model with a GCV score near the minimum is likely to have similar predictive power. *The flatness of the plot reflects the uncertainty in the data, and the resultant difficulty in choosing smoothing parameters.* We concluded earlier that $\alpha = 0.8$ was too large for the ethanol dataset; the lack of fit is reflected as the sharp increase in the GCV and CP scores at the left boundary of Figure 2.7. At the other end, we are unsure whether the additional features at $\alpha = 0.2$ in Figure 2.3 were real. The flat GCV plot reflects this uncertainty.

A consequence of Figure 2.7 is that going to extensive lengths to minimize GCV is very data-sensitive and can produce an unsatisfactory fit. In general, minimizing GCV (or CP, or CV) is highly variable: two visually similar datasets could produce very different results. Most importantly, just minimizing GCV discards significant information provided by the whole profile of the GCV curve, as displayed by the cross validation plot.

We should emphasize that the points raised here are *not* problems with cross validation and CP, but a reflection of the difficulty of model selection. This issue is explored further in Chapter 10, where cross validation methods are compared with bandwidth selectors claimed to be less variable. Such selectors are found to reflect the model selection difficulty in other ways; in particular, missing features when applied to difficult smoothing problems.

2.5   Linear Estimation

As noted previously, local regression is a linear estimate. The linear representation (2.12) provides the basis for a theoretical development of local regression estimation. Simple mean and variance expressions have already been derived; further properties are developed in this section. Results are also derived for the influence function and model selection criteria.

The first task is to identify the weight diagram. Let $\mathbf{X}$ be the $n \times (p+1)$ design matrix with rows $A(x_i - x)^T$, $\mathbf{W}$ be the diagonal matrix with entries

$w_i(x)$ and $Y = (Y_1, Y_2, \ldots, Y_n)^T$ be the response vector. The weighted sum of squares (2.5) can be written in matrix form

$$(Y - \mathbf{X}a)^T \mathbf{W}(Y - \mathbf{X}a).$$

If $\mathbf{WX}$ has full column rank, least squares theory gives the explicit expression

$$\hat{a} = (\mathbf{X}^T \mathbf{W}\mathbf{X})^{-1}\mathbf{X}^T \mathbf{W}Y \tag{2.23}$$

for the minimizer of (2.5).

The representation (2.23) identifies the weight diagram for the local polynomial smooth, defined by (2.12):

$$l(x)^T = e_1^T (\mathbf{X}^T \mathbf{W}\mathbf{X})^{-1}\mathbf{X}^T \mathbf{W}. \tag{2.24}$$

Here, e_1 is the unit vector; $e_1 = (1, 0, \ldots, 0)^T$.

The following theorem, originally from Henderson (1916) for local cubic fitting, provides a characterization of the weight diagrams for local polynomial regression.

Theorem 2.1 (Henderson's Theorem) The weight diagram for a local polynomial fit of degree p has the form

$$l_i(x) = W\left(\frac{x_i - x}{h(x)}\right) \langle \alpha, A(x_i - x) \rangle; \tag{2.25}$$

that is, the least squares weights multiplied by a polynomial of degree p. This representation is unique, provided $\mathbf{X}^T \mathbf{W}\mathbf{X}$ is non-singular.

Conversely, if a linear estimate reproduces polynomials of degree p, and the weight diagram has at most p sign changes, then the estimate can be represented as a local polynomial fit of degree p.

Proof: The representation (2.24) immediately yields (2.25), with $\alpha^T = e_1^T (\mathbf{X}^T \mathbf{W}\mathbf{X})^{-1}$, providing $\mathbf{X}^T \mathbf{W}\mathbf{X}$ is non-singular.

For the converse, define a polynomial $P(u - x)$ of degree $\leq p$, whose roots match the sign changes of the weight diagram. Then, the smoother is reconstructed as local polynomial smoothing with weights $w_i(x) = l_i(x)/P(x_i - x)$. $\qquad \square$

Despite its innocuous simplicity, Henderson's theorem has profound consequences. For example, the polynomial reproduction property implies the local regression method achieves exact bias correction in finite samples. The bias of a local linear estimate cannot depend on the slope $\mu'(x)$. See Section 2.5.2 for more discussion. As noted in Section 1.4, this contrasts sharply with kernel smoothing literature, where considerable effort has been expended in achieving asymptotic bias corrections.

For an immediate illustration of the power of Henderson's theorem, we derive a simplification of the leave-one-out cross validation statistic (2.20).

Theorem 2.2 If $\text{infl}(x_i) < 1$, the leave-one-out cross validation estimate $\hat{\mu}_{-i}(x)$ is

$$\hat{\mu}_{-i}(x_i) = \frac{\hat{\mu}(x_i) - \text{infl}(x_i)Y_i}{1 - \text{infl}(x_i)}$$

and

$$\text{CV}(\hat{\mu}) = \frac{1}{n}\sum_{i=1}^{n}\frac{(Y_i - \hat{\mu}(x_i))^2}{(1 - \text{infl}(x_i))^2}. \qquad (2.26)$$

This result can be proved directly using (2.12), (2.24) and some tedious matrix algebra. See Exercise 2.2 or Appendix B.4 of Myers (1990) for the same result for parametric regression. The following proof derives the result directly from Henderson's theorem.

Proof: Let

$$m_j(x_i) = \frac{l_j(x_i)}{1 - \text{infl}(x_i)}; j = 1, \ldots, n; j \neq i.$$

Using Henderson's theorem, we show that $\{m_j(x_i)\}$ is the weight diagram for $\hat{\mu}_{-i}(x_i)$ and thus

$$\hat{\mu}_{-i}(x_i) = \sum_{\substack{j=1 \\ j\neq i}}^{n} m_j(x_i)Y_j = \frac{\hat{\mu}(x_i) - \text{infl}(x_i)Y_i}{1 - \text{infl}(x_i)}.$$

(2.26) then follows directly from (2.20).

For fixed x_i, $\{m_j(x_i)\}$ is a polynomial multiplied by the weights $W((x_j - x_i)/h)$, because $\{l_j(x_i)\}$ is. It remains to show that $\{m_j(x_i)\}$ reproduces polynomials $P(x)$ of degree $\leq p$:

$$\sum_{\substack{j=1 \\ j\neq i}}^{n} m_j(x_i)P(x_j) = \frac{1}{1 - \text{infl}(x_i)}\left(\sum_{j=1}^{n} l_j(x_i)P(x_j) - l_i(x_i)P(x_i)\right)$$

$$= \frac{1}{1 - \text{infl}(x_i)}\left(P(x_i) - \text{infl}(x_i)P(x_i)\right)$$

$$= P(x_i)$$

where the second line follows from the polynomial reproducing property of $\{l_j(x_i)\}_{j=1}^{n}$. □

Theorem 2.2 assumes the bandwidth $h(x)$ does not change when the fit is carried out on the reduced dataset. This assumption can fail for a nearest neighbor bandwidth, but for estimating prediction error, this is the correct assumption to make, since the leave-one-out problem should mimic, as closely as possible, the true prediction problem.

The motivation for GCV also follows from the approximation (2.26), simply replacing $\text{infl}(x_i)$ by the average value $\text{tr}(\mathbf{L})/n$.

2.5.1 Influence, Variance and Degrees of Freedom

The variance of the local regression estimate was given by (2.13):

$$\text{var}(\hat{\mu}(x)) = \sigma^2 \|l(x)\|^2$$

or equivalently,

$$\text{var}(\hat{\mu}(x)) = \sigma^2 e_1^T (\mathbf{X}^T \mathbf{W} \mathbf{X})^{-1} (\mathbf{X}^T \mathbf{W}^2 \mathbf{X}) (\mathbf{X}^T \mathbf{W} \mathbf{X})^{-1} e_1. \qquad (2.27)$$

A quantity closely related to the variance is the influence function:

$$\text{infl}(x) = e_1^T (\mathbf{X}^T \mathbf{W} \mathbf{X})^{-1} e_1 W(0). \qquad (2.28)$$

This definition generalizes our earlier use of the influence function; the diagonal elements of the hat matrix $\mathbf{L}$ are given by $l_i(x_i) = \text{infl}(x_i)$.

Some important properties characterizing the relation between the influence function and variance are contained in the following theorem.

Theorem 2.3 Suppose the weight function $W(u)$ is non-negative, symmetric and decreasing on $[0, \infty)$. Then

1. the influence function dominates the variance;

$$\frac{1}{\sigma^2} \text{var}(\hat{\mu}(x)) \le \text{infl}(x). \qquad (2.29)$$

   For the rectangular weight function, this is an equality.

2. at the observation points x_i,

$$\text{infl}(x_i) \le 1 \qquad (2.30)$$

   and hence local regression is variance-reducing.

3. the influence function is a decreasing function of the bandwidths; i.e., if $h_1 < h_2$ and $\text{infl}_1(x)$ and $\text{infl}_2(x)$ are the corresponding influence functions, then

$$\text{infl}_1(x) \ge \text{infl}_2(x). \qquad (2.31)$$

Proof: Let α_i be the elements of $e_1^T (\mathbf{X}^T \mathbf{W} \mathbf{X})^{-1} \mathbf{X}^T$. Then

$$\frac{1}{\sigma^2} \text{var}(\hat{\mu}(x)) = \sum_{i=1}^{n} w_i(x)^2 \alpha_i^2 \le W(0) \sum_{i=1}^{n} w_i(x) \alpha_i^2 = \text{infl}(x)$$

since the weight function is maximized at 0. This proves (2.29). Exercise 2.2 establishes (2.30). Let $\mathbf{W}_1$ and $\mathbf{W}_2$ be the weight matrices corresponding to bandwidths h_1 and h_2. Since the weight function is decreasing on $[0, \infty)$,

$\mathbf{W}_1 \le \mathbf{W}_2$ on a componentwise basis and in particular, $\mathbf{X}^T(\mathbf{W}_2 - \mathbf{W}_1)\mathbf{X}$ is non-negative definite. Then

$$\text{infl}_2(x) = e_1^T(\mathbf{X}^T\mathbf{W}_1\mathbf{X} + \mathbf{X}^T(\mathbf{W}_2 - \mathbf{W}_1)\mathbf{X})^{-1}e_1 W(0) \ge \text{infl}_1(x),$$

using the results of Exercise 2.2. □

We gave two definitions for the degrees of freedom of the fit $\hat\mu(x)$ in (2.16). The residual degrees of freedom can be defined using similar techniques. Using

$$\text{var}(Y_i - \hat\mu(x_i)) = \sigma^2 - 2\text{cov}(Y_i, \hat\mu(x_i)) + \text{var}(\hat\mu(x_i)),$$

and a bias-variance decomposition, we have

$$\text{E}\sum_{i=1}^{n}(Y_i - \hat\mu(x_i))^2 = \sum_{i=1}^{n}\text{bias}(\hat\mu(x_i))^2 + \sigma^2(n - 2\nu_1 + \nu_2). \qquad (2.32)$$

This motivates the definition for residual degrees of freedom used in the variance estimate (2.18). It also establishes the unbiased estimation property for the CP statistic:

$$\text{ECP}(\hat\mu) = \frac{1}{\sigma^2}\sum_{i=1}^{n}\text{bias}(\hat\mu(x_i))^2 + \nu_2,$$

which is easily seen to equal $\text{E}(L(\hat\mu, \mu))$.

2.5.2   Bias

The bias of the linear estimate (2.12) is

$$\text{E}\hat\mu(x) - \mu(x) = \sum_{i=1}^{n}l_i(x)\mu(x_i) - \mu(x). \qquad (2.33)$$

Suppose $\hat\mu(x)$ is a local polynomial fit of degree p. Assuming that $\mu(x)$ is $p+2$ times differentiable, we can expand $\mu(\cdot)$ in a Taylor series around x:

$$\begin{aligned}
\mu(x_i) &= \mu(x) + (x_i - x)\mu'(x) + \ldots + (x_i - x)^p\frac{\mu^{(p)}(x)}{p!} \\
&\quad + (x_i - x)^{p+1}\frac{\mu^{(p+1)}(x)}{(p+1)!} + (x_i - x)^{p+2}\frac{\mu^{(p+2)}(x)}{(p+2)!} + \ldots.
\end{aligned}$$

As an application of Henderson's theorem, we know $\sum_{i=1}^{n}l_i(x)(x_i-x)^j = 0$ for $1 \le j \le p$. This leads to

$$\begin{aligned}
\text{E}(\hat\mu(x)) - \mu(x) &= \frac{\mu^{(p+1)}(x)}{(p+1)!}\sum_{i=1}^{n}l_i(x)(x_i - x)^{p+1} \\
&\quad + \frac{\mu^{(p+2)}(x)}{(p+2)!}\sum_{i=1}^{n}l_i(x)(x_i - x)^{p+2} + \ldots. \qquad (2.34)
\end{aligned}$$

The bias has a leading term involving the $(p+1)$st derivative $\hat{\mu}^{(p+1)}(x)$. This is very similar to the systematic error for graduation rules discussed in Section 1.1.2. There, the systematic error of cubic reproducing rules was dominated by the fourth differences of the mean function.

Why keep the $\mu^{(p+2)}(x)$ term in (2.34)? Suppose the design points are equally spaced. In this case, the weight diagram is symmetric around the fitting point x. If p is even, $p+1$ is odd and $\sum_{i=1}^{n} l_i(x)(x_i - x)^{p+1} = 0$ by symmetry. Thus, the first term in the bias expansion disappears. In this case, the second term of (2.34) is dominant. For nonuniform data and even p, the $\mu^{(p+1)}$ and $\mu^{(p+2)}$ terms are generally of similar size.

For multidimensional predictors x_i, an expression similar to (2.34) still holds, but it will be the sum of terms involving *all* partial derivatives of $\mu(x)$ of orders $p+1$ and $p+2$.

2.6   Asymptotic Approximations

Explicit closed form expressions such as (2.27) are suitable for numerical computation. However, they only give limited insight into the behavior of the variance function as, for example, the design, sample size or bandwidth change. In this section we provide some simple asymptotic approximations to the bias, variance and influence functions. Similar results, particularly for the bias and variance, have been derived in many places for kernel estimates. Work on local regression includes Lejeune (1985), Tsybakov (1986), Müller (1987), Ruppert and Wand (1994) and Fan and Gijbels (1996).

To state asymptotic results, we need to make assumptions about how the sequence of design points $x_1, \ldots, x_n$ behaves as n increases. A common assumption is **random design**: the points are sampled from a continuous density $f(x)$. Another model is the **regular design** (often called the fixed design); this includes the common case of equally spaced points. More generally, a regular design generated by a density $f(u)$ defines $x_{i,n}$ to be the solution of

$$\frac{i - 0.5}{n} = \int_{-\infty}^{x_{i,n}} f(u)du.$$

The matrix $\mathbf{X}^T\mathbf{W}\mathbf{X}$ has components of the form $\sum_{i=1}^{n} w_i(x)(x_i - x)^k$. Under either the random or regular design, and mild regularity conditions (in particular, $nh^d \to \infty$):

$$\frac{1}{nh^d} \sum_{i=1}^{n} w_i(x)^j \frac{(x_i - x)^k}{h^k} = \int W(v)^j v^k f(x + hv)dv + o(1). \qquad (2.35)$$

This result is valid for fixed h. Asymptotic results are often stated under the additional assumption $h \to 0$, for which (2.35) simplifies to

$$\frac{1}{nh^d} \sum_{i=1}^{n} w_i(x)^j \frac{(x_i - x)^k}{h^k} \to f(x) \int W(v)^j v^k dv. \qquad (2.36)$$

If one requires consistency of the local regression estimate, small bandwidth results are required. But Stoker (1993) argues that results based on fixed bandwidth asymptotics provide a better characterization of results in finite samples. In particular, one should use results based on the fit actually and not on the basis of what one promises to do if the sample size were larger.

For regular designs, the limit (2.36) follows from the theory of Reimann sums. For random design, the result is consistency for a kernel density estimate; see Chapter 5 (strong convergence requires slightly stronger bandwidth conditions). More generally, we can take (2.36) as a basic assumption: Results that follow from (2.36) will hold for regular, random and any other design satisfying this condition.

Applying (2.35) and (2.36) to the matrix $\mathbf{X}^T \mathbf{W}^j \mathbf{X}$ gives

$$\frac{1}{nh^d} \mathbf{H}^{-1} \mathbf{X}^T \mathbf{W}^j \mathbf{X} \mathbf{H}^{-1}$$
$$= \begin{cases} \int \int W(v)^j A(v) A(v)^T f(x + hv) dv + o(1), & h \text{ fixed} \\ f(x) \int W(v)^j A(v) A(v)^T dv + o(1) & h \to 0 \end{cases} \qquad (2.37)$$

where $\mathbf{H}$ is a diagonal matrix with elements $1, h, \ldots, h^p$. Asymptotic approximations to quantities such as the bias and variance are now easily derived.

Variance and Influence Functions. Under the small bandwidth limits, the influence function (2.28) and variance (2.27) have asymptotic approximations

$$\text{infl}(x) = \frac{W(0)}{nh^d f(x)} e_1^T \mathbf{M}_1^{-1} e_1 + o((nh)^{-1}) \qquad (2.38)$$

$$\text{var}(\hat{\mu}(x)) = \frac{\sigma^2}{nh^d f(x)} e_1^T \mathbf{M}_1^{-1} \mathbf{M}_2 \mathbf{M}_1^{-1} e_1 + o((nh)^{-1}) \qquad (2.39)$$

where $\mathbf{M}_j = \int W(v)^j A(v) A(v)^T dv$. Similar expressions are easily derived for fixed bandwidth limits. The influence function and variance are inversely proportional to the number of data points in the smoothing window, with the proportionality constant depending on the weight function, degree of fit and σ^2.

The asymptotic approximations (2.38) and (2.39) are mainly of theoretical interest; they should never be considered an alternative to (2.28) and (2.27) for actually computing the influence and variance. First, the approximations introduce another unknown, $f(x)$, to estimate. Second, the approximations can be exceedingly poor in the tails. Third, the finite sample versions are cheap to compute; see Section 12.3. Exercise 3.5 provides some comparisons.

Asymptotic Equivalent Kernels. Substituting (2.37) into the expression (2.23) for the local regression estimate yields

$$\hat{\mu}(x) \approx \frac{1}{nhf(x)} e_1^T \mathbf{M}_1^{-1} \mathbf{H}^{-1} \mathbf{X}^T \mathbf{W} \mathbf{Y}$$

$$= \frac{1}{nhf(x)} \sum_{i=1}^{n} W^* \left(\frac{x_i - x}{h} \right) Y_i$$

where

$$W^*(v) = e_1^T \mathbf{M}_1^{-1} A(v) W(v). \tag{2.40}$$

The weight function $W^*(v)$, dependent on the degree of fit and the original weight function $W(v)$, is the asymptotically equivalent kernel. Roughly, for large samples, the local regression estimate is equivalent to a local constant estimate using the weight function $W^*(v)$. Equivalent kernels often provide poor approximations, But they simplify theoretical computations considerably. The asymptotic variance (2.39) is simply

$$\text{var}(\hat{\mu}(x)) \approx \frac{\sigma^2}{nhf(x)} \int W^*(v)^2 dv.$$

The influence function approximation is even simpler:

$$\text{infl}(x) \approx \frac{W^*(0)}{nhf(x)}.$$

Bias. The first term of the bias expansion (2.34) is approximated by

$$b(x) = \frac{h^{p+1} \mu^{(p+1)}(x)}{(p+1)!} \int v^{p+1} W^*(v) dv + o(h^{p+1}). \tag{2.41}$$

If p is even and $W(v)$ is symmetric, $\int v^{p+1} W^*(v) dv = 0$. The dominant bias arises from the second term of (2.34), which has size $O(h^{p+2})$. But care is needed, since there are other terms of the same size. For p even, one obtains

$$b(x) = h^{p+2} \left(\frac{\mu^{(p+2)}(x)}{(p+2)!} + \frac{\mu^{(p+1)}(x) f'(x)}{(p+1)! f(x)} \right) \int v^{p+2} W^*(v) dv + o(h^{p+2}).$$

See Ruppert and Wand (1994) for a full derivation.

Degrees of Freedom. Suppose the design is uniform on $[0, 1]$, so that $f(x) = 1$ for $0 \le x \le 1$. Then

$$\text{tr}(\mathbf{L}) = \sum_{i=1}^{n} \text{infl}(x_i) \approx \frac{W^*(0)}{h} \tag{2.42}$$

Degree	Constant h	Nearest Neighbor
0	$0.0617 + 0.8198\nu_1$	$0.0465 + 0.8198\nu_1$
1	$0.2978 + 0.8198\nu_1$	$0.2493 + 0.8198\nu_1$
2	$0.1702 + 0.8965\nu_1$	$0.1502 + 0.8965\nu_1$

TABLE 2.1. Approximate relation between influence degrees of freedom ν_1 and variance degrees of freedom ν_2 using the tricube weight function. Results are shown for both constant and nearest neighbor bandwidths.

and

$$\text{tr}(\mathbf{L}^T\mathbf{L}) \approx \frac{1}{h}\int W^*(v)^2 dv.$$

These expressions are dependent on the uniform design. In fact, a more detailed analysis shows that $\text{tr}(\mathbf{L})$ and $\text{tr}(\mathbf{L}^T\mathbf{L})$ have limits that are linear in $1/h$; the second term arising from boundary effects. Eliminating h gives

$$\text{tr}(\mathbf{L}^T\mathbf{L}) \approx \frac{\int W^*(v)^2 dv}{W^*(0)}\text{tr}(\mathbf{L}) + c \qquad (2.43)$$

for a constant c. This expression has no dependence on the bandwidth. Moreover, the approximation holds fairly universally, irrespective of the design. To evaluate the constant c, one has to use fixed bandwidth limits based on (2.35), carefully treating boundary effects. An alternative is to estimate the constant by evaluating ν_1 and ν_2 in a few cases.

Table 2.1 summarizes the approximate relationships between $\nu_1 = \text{tr}(\mathbf{L})$ and $\nu_2 = \text{tr}(\mathbf{L}^T\mathbf{L})$ for the tricube weight function. In all cases, the linear component dominates the constant component. As expected, the linear coefficient is always slightly less than 1.

Bandwidths. Consider the one dimensional local linear regression, and suppose the design density $f(x)$ has unbounded support. Then (2.39) reduces to

$$\text{var}(\hat{\mu}(x)) = \frac{\sigma^2}{nhf(x)}\frac{\int W(v)^2 dv}{(\int W(v)dv)^2} + o((nh)^{-1}) \qquad (2.44)$$

while (2.41) reduces to

$$\text{E}(\hat{\mu}(x)) = \mu(x) + \frac{h^2\mu''(x)}{2}\frac{\int v^2 W(v)dv}{\int W(v)dv} + o(h^2). \qquad (2.45)$$

Clearly, as h increases with n fixed, the variance decreases while the bias increases. This is a mathematical demonstration of the bias-variance trade-off. Combining these yields the mean squared error approximation

$$\text{E}((\hat{\mu}(x) - \mu(x))^2) \approx \frac{\sigma^2}{nhf(x)}\frac{\int W(v)^2 dv}{(\int W(v)dv)^2} + \frac{h^4\mu''(x)^2}{4}\left(\frac{\int v^2 W(v)dv}{\int W(v)dv}\right)^2.$$

Minimizing over h yields the asymptotically optimal bandwidth

$$h_{opt}^5 = \frac{\sigma^2 \int W(v)^2 dv}{nf(x)\mu''(x)^2 \left(\int v^2 W(v)dv\right)^2}. \tag{2.46}$$

This result is not useful in selecting the bandwidth, since it depends on unknown quantities; in particular, $\mu''(x)$. Substituting estimates for $\mu''(x)$ isn't helpful, since it doesn't address the main question for bandwidth and model selection: What features are real? This is discussed further in Chapter 10.

Asymptotic Normality. If the errors ϵ_i are normally distributed with variance σ^2, then

$$\frac{\hat{\mu}(x) - E(\mu(x))}{\sigma \|l(x)\|} \tag{2.47}$$

has a $N(0,1)$ distribution. If the errors are independent and identically distributed with finite variance σ^2, but with a non-normal distribution, one can easily check that the Lindeberg condition (Shiryayev 1984, page 326) holds, provided $\max_{1 \le i \le n} |l_i(x)|/\|l(x)\| \to 0$. Thus, (2.47) has an asymptotic $N(0,1)$ distribution. Under both random and regular designs, the conditions $nh^d \to \infty$ and $f(x) > 0$ are sufficient.

Data-driven Bandwidths. The results stated in this section have all assumed that the bandwidth sequence $h = h_n$ is nonrandom, and in particular, is independent of the responses Y_i. To extend results to a data-driven sequence $h = \hat{h}_n$ (for example, chosen by cross validation), one identifies a deterministic sequence $\tilde{h}_n$ such that $\hat{h}_n/\tilde{h}_n \to 1$ in probability. Results of this type are found in Rice (1984) and elsewhere. Then, one shows that smoothing with $\hat{h}_n$ is asymptotically equivalent to smoothing with $\tilde{h}_n$.

2.7 Exercises

2.1 Derive the closed form (2.8) for the local linear estimate. It may be easier to fit the local model $b_0 + b_1(x_i - \bar{x}_w)$, and $\hat{\mu}(x) = \hat{b}_0 + \hat{b}_1(x - \bar{x}_w)$. Convince yourself this must produce the same answer as the local linear estimate defined by (2.6).

2.2   a) Let $\mathbf{V}$ be an invertible $p \times p$ matrix; a a p-vector and λ a constant with $1 + \lambda a^T \mathbf{V}^{-1} a \ne 0$. Show

$$\left(\mathbf{V} + \lambda a a^T\right)^{-1} = \mathbf{V}^{-1} - \frac{\lambda \mathbf{V}^{-1} a a^T \mathbf{V}^{-1}}{1 + \lambda a^T \mathbf{V}^{-1} a}. \tag{2.48}$$

Use this result to provide a direct matrix proof of Theorem 2.2.

b) Show that

$$a^T \left(\mathbf{V} + \lambda a a^T\right)^{-1} a = \frac{\lambda a^T \mathbf{V}^{-1} a}{1 + \lambda a^T \mathbf{V}^{-1} a}.$$

Hence prove (2.30).

c) Suppose $\mathbf{V}$ and Σ are non-negative definite. Show that

$$a^T(\mathbf{V}+\Sigma)^{-1}a \geq a^T\mathbf{V}^{-1}a.$$

2.3 a) Let $\mathbf{V}$, a and λ be as in Exercise 2.2, with $d = \lambda - a^T\mathbf{V}a \neq 0$. Show

$$\begin{pmatrix} \mathbf{V} & a \\ a^T & \lambda \end{pmatrix}^{-1} = \begin{pmatrix} \mathbf{V}^{-1} + \frac{1}{d}\mathbf{V}^{-1}aa^T\mathbf{V}^{-1} & -\frac{1}{d}\mathbf{V}^{-1}a \\ -\frac{1}{d}a^T\mathbf{V}^{-1} & \frac{1}{d} \end{pmatrix}.$$

b) Show the influence function $\mathrm{infl}(x)$ increases as the degree of the local polynomial increases.

c) Provide an example to show that local linear fitting can be less variable than local constant fitting.

2.4 Consider the local constant estimate (2.7).

a) Write down the variance of this estimate, assuming the Y_i are independent and have constant variance.

b) Suppose the weight function $W(u)$ is symmetric, non-negative and $uW'(u)/W(u)$ is monotone decreasing on $[0,\infty)$. Show, for fixed x, that the variance is a decreasing function of the bandwidth h.

c) Relaxing the conditions on the weight function, provide an example to show the variance can increase as the bandwidth h increases.

2.5 *Double Smoothing*. The local regression estimate $\hat{\mu} = \mathbf{L}Y$ has bias $b = \mathrm{E}(\hat{\mu}) - \mu = -(\mathbf{I} - \mathbf{L})\mu$. Double smoothing estimates the bias by $\hat{b} = -(\mathbf{I} - \mathbf{L})\hat{\mu}$, producing the bias corrected estimate $\hat{\hat{\mu}} = \hat{\mu} - \hat{b} = (2\mathbf{I} - \mathbf{L})\mathbf{L}Y$. Suppose the design is equally spaced with $x_i = i; i = \ldots, -2, -1, 0, 1, 2, \ldots$; the doubly infinite sequence avoids boundary effects.

a) Suppose the smoother is a moving average (local constant) fit with bandwidth h;

$$\hat{\mu}(i) = \frac{1}{2h+1}\sum_{j=-h}^{h} Y_{i+j}.$$

Show that

$$\hat{\hat{\mu}}(i) = \frac{1}{2h+1}\sum_{j=-2h}^{2h}\left(2I(|j| \leq h) - 1 + \frac{|j|}{2h+1}\right)Y_{i+j}.$$

b) Show that the weight diagram for $\hat{\mu}(i)$ is quadratic reproducing, and hence by Henderson's theorem $\hat{\mu}$ is a local quadratic smooth. Does the weight diagram look sensible?

2.6 Some visual experiments:

a) Construct a dataset with a mean function having flat and steep regions. For example, let x_i be uniform on the interval $[-5, 5]$ and $Y_i = \Phi(x_i) + \epsilon_i$ where $\Phi(x)$ is the standard normal distribution function and ϵ_i is normally distributed with $\sigma = 0.1$. Plot the dataset. Does the residual variance *look* constant?

b) Construct a nonuniform predictor variable. For example, in S-Plus, x <- sqrt(runif(100)). Generate standard normal observations as the response variable. Plot the data. Does the residual variance *look* constant? This experiment may take two or three attempts; eventually, large residuals in the high density region should be distracting.

Remark. A real data example where these visual distractions occur is provided in Exercise 3.2.

2.7 *Estimation under the L_1 loss function.*

a) Suppose $X \sim N(\mu, \sigma^2)$. Show that

$$E|X| = \mu \left(\Phi(\frac{\mu}{\sigma}) - \Phi(-\frac{\mu}{\sigma}) \right) + 2\sigma\phi(\frac{\mu}{\sigma}).$$

Here, $\phi(\cdot)$ and $\Phi(\cdot)$ denote the standard normal density and distribution function.

b) Consider the risk function $R_1(\hat{\mu}, \mu) = n^{-1} \sum_{i=1}^{n} |\hat{\mu}(x_i) - \mu(x_i)|$. Derive an explicit expression for $R_1(\hat{\mu}, \mu)$ in terms of the bias and variance.

2.8 L_1 *Cross Validation.* Suppose the risk function for predicting a future observation (x_{new}, Y_{new}) is $E|Y_{new} - \hat{\mu}(x_{new})|$. This is estimated by the L_1 cross validation criterion

$$CV_1(\hat{\mu}) = \frac{1}{n} \sum_{i=1}^{n} |Y_i - \hat{\mu}_{-i}(x_i)|.$$

Show that

$$CV_1(\hat{\mu}) = \frac{1}{n} \sum_{i=1}^{n} \frac{|Y_i - \hat{\mu}(x_i)|}{1 - \text{infl}(x_i)}.$$

Propose an L_1 version of generalized cross validation.

3
Fitting with LOCFIT

The examples in this book are implemented using the local regression software LOCFIT. This can be used either as a stand-alone program or as a library within the S (Becker, Chambers and Wilks 1988), S-Plus or R (Ihaka and Gentleman 1996) languages. See Appendix A for details of how to obtain the LOCFIT code and installation.

The code examples in this book are designed for S version 4; most will also work in S-Plus and R. The syntax for the stand-alone C-LOCFIT version is different. For many examples, the corresponding code for the stand-alone version can be obtained using the **example** command:

```
locfit> example 3.1
```

```
Example 3.1. Local Regression
```

```
  locfit NOx~E data=ethanol alpha=0.5
  plotfit data=T
```

prints the corresponding code on the screen. Typing

```
locfit> example 3.1 run
```

results in the code being executed and plots being produced as appropriate.

3.1 Local Regression with LOCFIT

LOCFIT provides two functions, `locfit()` and `locfit.raw()`, to perform local regression. The `locfit()` function uses the S model language to specify the local regression model, while `locfit.raw()` has separate arguments for the predictor and response variables. In other respects, the two functions are identical.

Example 3.1. We fit a local quadratic model to the ethanol dataset and plot the result:

```
> fit <- locfit(NOx~E, data=ethanol, alpha=0.5)
> fit
Call:
locfit(formula = NOx ~ E, data = ethanol, alpha = 0.5)

Number of observations:          88
Fitted Degrees of freedom:       6.485
Residual scale:                  0.336
> plot(fit, get.data=T)
```

The plot was displayed in Figure 2.2.

The first argument to `locfit()` is the *model formula* `NOx~E` specifying the local regression model and is read as "NOx is modeled by E". The `data=ethanol` argument specifies a data frame where the variables in the model formula may be found; if the `data` argument is omitted, currently attached data directories are searched. The use of model formulae and data frames follows chapter 2 of Chambers and Hastie (1992). The third argument to the `locfit()` function, `alpha`, controls the bandwidth. Here, a nearest neighbor based bandwidth covering 50% of the data is used. The fit could also be generated by

```
> fit <- locfit.raw(ethanol$E, ethanol$NOx, alpha=0.5)
```

The fit returned by the `locfit()` call is an S object, with the `"locfit"` class. Printing the fit then shows a short summary. The `plot(fit)` command then calls the plot method `plot.locfit()`. The `get.data=T` argument adds the original data to the plot.

Confidence intervals can be added to the plot with the `band=` argument, for example,

```
> plot(fit,band="global")
```

adds confidence intervals under the assumption that the residual variance σ^2 is constant. If `band="local"`, an attempt is made to estimate σ^2 locally. If `band="pred"`, prediction bands (2.17) are computed under the constant variance assumption. Variance estimation and confidence bands are discussed in more detail in Chapter 9.

3.2 Customizing the Local Fit

The `locfit()` function has additional arguments to control the fit. The most important are described in this section; others are introduced throughout the book as they are needed.

Smoothing Parameter. The `alpha` argument, used in Example 3.1, controls the bandwidth. When `alpha` is given as a single number, it represents a *nearest neighbor* fraction, as described in section 2.2.1.

Example 3.2. (Changing the Smoothing Parameter.) We compute local regression fits for the ethanol dataset, with four different smoothing parameters:

```
> alp <- c(0.8,0.6,0.4,0.2)
> for(a in alp) {
+   fit <- locfit(NOx~E, data=ethanol, alpha=a)
+   plot(fit, get.data=T, main=paste("alpha =",a))
+ }
```

The fits are as shown in Figure 2.3 (For the actual code producing the trellis display, see section B.4).

More generally, `alpha` can be specified as a vector with two components. The second component represents a constant bandwidth, so `alpha=c(0,1)` implies $h(x) = 1$ is used everywhere. If both the nearest neighbor and fixed components are nonzero, both bandwidths are computed, and $h(x)$ will be chosen as the larger component. Specifically, if $\alpha = (\alpha_0, \alpha_1)$, the bandwidth $h(x)$ is computed as follows:

1. $k = \lfloor n\alpha_0 \rfloor$.

2. Compute $d_i = |x - x_i|; i = 1, \ldots, n$ and find the kth smallest $d_{(k)}$.

3. Return $h(x) = \max(d_{(k)}, \alpha_1)$.

The default smoothing parameter is `alpha=c(0.7,0)`.

Degree of Local Polynomial. The degree of local polynomial is specified through the `deg` argument: `deg=1` specifies a local linear fit; `deg=2` species local quadratic (the default). For univariate fits, LOCFIT supports any degree, although there's usually little reason to use degrees greater than 3. For multivariate fits, `deg=3` is the maximum.

The Weight Function. LOCFIT supports several weight functions, listed in Table 3.1. These are selected with the `kern` argument to the `locfit()` function. For example,

```
> locfit(...,kern="gauss")
```

selects the Gaussian weight function. The default weight function is the tricube. Note also the factor of 2.5 in the Gaussian weight function; this makes the scaling for the Gaussian weight function more comparable to the compact weight functions.

rect	Rectangular	$W(x) = 1, |x| < 1$
tria	Triangular	$W(x) = 1 - |x|, |x| < 1$
epan	Epanechnikov	$W(x) = 1 - x^2, |x| < 1$
bisq	Bisquare	$W(x) = (1 - x^2)^2, |x| < 1$
tcub	Tricube	$W(x) = (1 - |x|^3)^3, |x| < 1$
trwt	Triweight	$W(x) = (1 - x^2)^3, |x| < 1$
gauss	Gaussian	$W(x) = \exp(-(2.5x)^2/2)$
expl	Exponential	$W(x) = \exp(-3|x|)$
minm	Minimax	See Section 13.3
macl	McLain	$W(x) = 1/(x + \epsilon)^2$

TABLE 3.1. The LOCFIT weight functions.

3.3   The Computational Model

The definition of local regression formally requires solving a weighted least squares problem for each fitting point x. But for large datasets, or the iterative procedures discussed in later chapters, this becomes computationally expensive.

The idea of a computational model began with the LOWESS algorithm (Cleveland 1979) and was developed considerably by LOESS (Cleveland and Grosse 1991). The local regression method is carried out at a small set of fitting points. The fitted values and local slopes at these fitting points are then used to define a fitted surface, which can be evaluated rapidly at any point. LOCFIT uses a similar computational model but differs in the way the fitting points are chosen. In particular, the LOCFIT computational model is bandwidth adaptive, choosing the most fitting points in regions where the smallest bandwidths are used. The algorithm is described more fully in Chapter 12.

The determination of fitting points and the direct fitting are performed by the locfit() function. The predict.locfit() and preplot.locfit() methods are used to interpolate the fits. These functions have a similar set of arguments but differ in the returned objects: predict.locfit() returns a vector of the predicted values, while preplot.locfit() returns an object with the "preplot.locfit" class. This object contains prediction points, predicted values and other information required to produce a plot.

Example 3.3. For the fit to the ethanol dataset, the fitted surface is evaluated at $E = 0.6, 0.8$ and 1.0:

```
> fit <- locfit(NOx~E, data=ethanol, alpha=0.5)
```

```
> predict(fit, c(0.6,0.8,1.0))
[1] 0.7239408 2.7544413 3.1183651
```

The two arguments to predict() are the "locfit" object and the
newdata of prediction points. The latter can have one of several forms,
including a vector, data frame, matrix or grid margins. See Appendix B.1
for more details.

3.4 Diagnostics

3.4.1 Residuals

The residuals of a LOCFIT model can be found with the command

```
> res <- residuals(fit)
```

which calls the residuals method residuals.locfit().

Example 3.4. Smoothed residual plots are constructed for the four fits
in Figure 2.3:

```
> alp <- c(0.8,0.6,0.4,0.2)
> for(a in alp) {
+    fit <- locfit(NOx~E, data=ethanol, alpha=a)
+    res <- residuals(fit)
+    fit2 <- locfit.raw(ethanol$E, res, alpha=0.2)
+    plot(ethanol$E, res, main=paste("alpha =",a),
+      ylim = c(-1,1))
+    lines(fit2)
+    abline(h=0, lty=2)
+ }
```

Figure 2.5 showed the smoothed residual plots. Note that locfit.raw()
is used to smooth the residuals, since the residuals are not stored on the
ethanol data frame.

3.4.2 Cross Validation

The cross validation and CP criteria can be computed from information
stored on a "locfit" object. We begin with GCV, since this is most direct.

Example 3.5. From the ethanol fit, we extract a dp component that
contains information about the fit:[1]

```
> fit <- locfit(NOx~E, data=ethanol, alpha=0.5)
```

---

[1] Here, and elsewhere, users of S version 3 must substitute $ for @.

```
> fit@dp
 nnalph fixh adpen cut      lk       df1       df2        rv
   0.5    0     0 0.8 -4.53376 7.013307 6.487448 0.1126948
```

The components of interest to us are lk (-0.5 times the residual sum of squares), df1 ($tr(\mathbf{L})$) and df2 ($tr(\mathbf{L'L})$). Since this dataset contains 88 points, the GCV score is computed as

```
> gencv <- 88*(-2*fit@dp["lk"])/(88-fit@dp["df1"])^2
> gencv
       lk
 0.1216522
```

In fact, LOCFIT provides two functions, gcv() and gcvplot(), to simplify this. gcv() automatically calls locfit(), and returns a vector with four components: the log-likelihood (-0.5 times the residual sum of squares), the degrees of freedom according to the influence and variance definitions, and the GCV score. The arguments for gcv() are exactly the same as for locfit().

gcvplot() is a wrapper function for gcv(). It is provided a vector of smoothing parameters, and calls gcv() in turn for each parameter. It returns an object with the "gcvplot" class; the plot method defined for this class produces cross validation plots such as those in Figure 2.7.

Example 3.6. The gcvplot() function is applied to the ethanol dataset for a range of smoothing parameters:

```
> alpha <- seq(0.2, 0.8, by=0.05)
> plot(gcvplot(NOx~E, data=ethanol, alpha=alpha),
+    ylim=c(0,0.2))
```

Figure 2.7 showed the result. Note that each smoothing parameter here is a nearest neighbor fraction; to use constant bandwidths, alpha should be a two-column matrix with the first column 0.

The cross validation approach is only slightly more complicated. We can use the definition directly by using a special cross validation evaluation structure, ev="cross":

```
> fit <- locfit(NOx~E, data=ethanol, alpha=0.5, ev="cross")
> -2*fit@dp["lk"]/88
       lk
 0.1171337
```

This deletes each observation in turn and computes the fit, so should only be used for fairly small datasets. For large datasets, an approximation is

```
> fit <- locfit(NOx~E, data=ethanol, alpha=0.5)
> r <- residuals(fit)
```

```
> infl <- fitted(fit,what="infl")
> mean((r/(1-infl))^2)
[1] 0.1190185
```

The small discrepancy here is because the fitted values and influence function are being interpolated rather than computed directly at each point. A simpler alternative is

```
> mean(residuals(fit,cv=T)^2)
[1] 0.1177989
```

When provided with the cv=T argument, the residuals.locfit() function computes the values

$$(1 + \text{infl}(x_i))(Y_i - \hat{\mu}(x_i)). \tag{3.1}$$

Thus the sum of squares in this case is

$$\sum_{i=1}^{n} ((1 + \text{infl}(x_i))(Y_i - \hat{\mu}(x_i)))^2 \tag{3.2}$$

rather than the exact cross validation. Clearly, the two approaches are asymptotically equivalent in large samples, when $\text{infl}(x_i)$ is small. The motivation for (3.1) will become clear in Chapter 4, where (3.1) generalizes naturally to local likelihood problems. Droge (1996) argued that (3.2) provides a better estimate of the prediction mean squared error in finite samples. A pair of functions, lcv() and lcvplot(), are provided to implement this cross validation method.

The pair of functions, cp() and cpplot(), implement the CP method. The implementation is again similar to gcv(), but now requires an estimate of the residual variance σ^2. By default, cpplot() takes the variance estimate (2.18) from the fit with the largest degrees of freedom ν_2.

3.5 Multivariate Fitting and Visualization

To specify a multivariate local regression model, multiple terms are specified on the right-hand side of the model formula.

Example 3.7. We consider the ethanol dataset used in Figure 2.1. A second predictor variable, C, was not considered previously and measures the compression ratio of the engine. The fit is computed by:

```
> fit <- locfit(NOx~E+C, data=ethanol, alpha=0.5, scale=0)
> plot(fit, get.data=T)
> plot(fit, type="persp")
```

The formula can be given either as NOx~E+C or NOx~E*C; both will give the same results. Figure 3.1 shows the resulting contour and perspective plots. If type="image", the plot is produced using the S-Plus image() function.

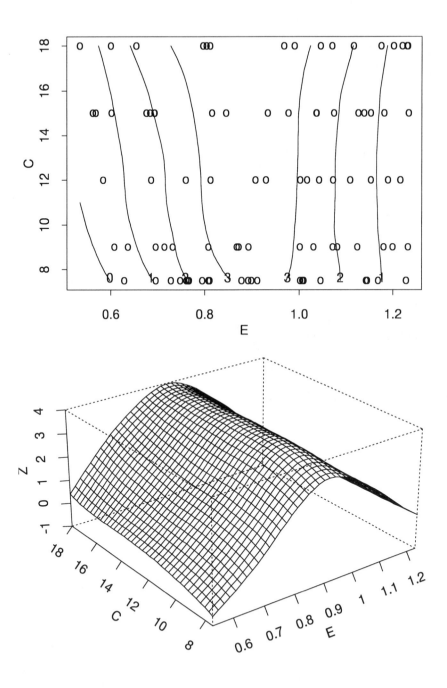

FIGURE 3.1. Bivariate local regression for the ethanol data.

An important argument in the multivariate case is `scale`. This provides a set of scales s_j to use in the distance computation (2.10), thereby controlling the relative amounts of smoothing in each variable. Specifying `scale=0` results in the sample standard deviations of each of the variables being computed and used as scales. One can also compare fits with different choices of scales using cross validation methods.

The contour plot and perspective plot in Figure 3.1 both display the fitted surface, but serve different visualization purposes. No general claim can be made that one of these displays is better. Either display shows that NOx is much more heavily dependent on E than on C. The general nature of this dependence - first increasing, then decreasing, as E increases - is more readily apparent in the perspective plot. On the other hand, values of the fitted surface are more readily judged from the contour plot. As an exercise, try to estimate $\hat{\mu}(0.7, 13)$ from the perspective plot, and then do the same from the contour plot.

What about the dependence on C? Both plots appear to show some increase in NOx with C, although it is difficult to preceive the precise nature of the relationship. Is there any interaction between the two variables? Neither plot is good for answering this type of question. An alternative display is of one dimensional cross sections of the fit. The S Trellis library (Becker, Cleveland, Shyu and Kaluzny 1994) provides a convenient mechanism for producing such plots. An interface is provided in the `plot.locfit()` function, by specifying a *panel variable* pv, which is varied within a panel of the trellis display, and a `tv`, which is varied between panels of the trellis display. A final argument, `mtv`, specifies the number of panels for the display.

Example 3.8. To run this example, trellis graphics must be initialized, using the `trellis.device()` function. We plot the bivariate fit to the ethanol dataset, using E as the trellis variable:

```
> fit <- locfit(NOx~E+C, data=ethanol, alpha=0.5, scale=0)
> plot(fit, pv="C", tv="E", mtv=9, get.data=T)
```

Figure 3.2 display the results. The slight dependence of NOx on C is much easier to see in this plot than in Figures 3.1 and 3.3.

3.5.1 Additive Models

The definition of multivariate local regression extends to any number of dimensions. But beyond two or three dimensions, a local regression model is difficult to fit, due to both the rapid increase in the number of parameters in the local model and the sparsity of data in high dimensional spaces. In addition, visualization of a high dimensional surface is difficult.

Because of these problems, a number of simplified models have been proposed. Typically, these methods build a fitted surface by applying local regression (or other smoothers) to low dimensional projections of the data.

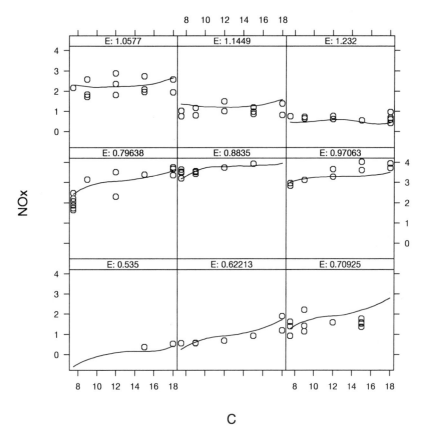

FIGURE 3.2. Bivariate local regression: Ethanol data sectioned by equivalence ratio E.

The most widely studied model of this type is the additive model, which for two predictors x and z has the form

$$\mu(x, z) = \mu_1(x) + \mu_2(z)$$

where $\mu_1(x)$ and $\mu_2(z)$ are smooth functions. The backfitting algorithm can be used to fit the model and alternately estimates the components. A thorough account of additive models and the backfitting algorithm can be found in Hastie and Tibshirani (1990). Opsomer and Ruppert (1997) discuss some theoretical properties of the backfitting algorithm.

Additive models are fitted in S using the gam() function described in Hastie (1992). To use LOCFIT for the additive component, functions lf() and gam.lf() are provided. The lf() function is used in the model formula; at the time of writing it accepts the alpha, deg, ev and kern arguments. We remark that the gam library also includes a lo() function for fitting additive terms using local regression and LOESS. The `lf()` function has considerably more flexibility.

Example 3.9. An additive model is fitted to the ethanol dataset. We use a local quadratic term with `alpha=0.5` for the equivalence ratio and a local linear term for the compression ratio.

```
> library("locfit",first=T)
> fit <- gam(NOx~lf(E,alpha=0.5)+lf(C,deg=1),data=ethanol)
> plot(fit)
```

Figure 3.3 plots the two additive components, showing nearly linear dependence on C and the peaked dependence on E. One has to look closely at the scales to see that the E dependence is much stronger.

Important: For this example to work properly, you *must* specify `first=T` when attaching the LOCFIT library, or otherwise ensure `"lf"` appears in your `gam.slist` variable.

A special case of the additive model is the partially linear model:

$$\mu(x, z) = \mu_1(x) + \langle \beta, z \rangle .$$

This model is particularly attractive since the backfitting algorithm has a closed form limit:

$$\hat{\beta} = \left(\mathbf{X}_2^T (\mathbf{I} - \mathbf{L}_1) \mathbf{X}_2 \right)^{-1} \mathbf{X}_2^T (\mathbf{I} - \mathbf{L}_1) Y, \tag{3.3}$$

where $\mathbf{X}_2$ is the design matrix for the parametric component and $\mathbf{L}_1$ is the hat matrix for the smooth component. See Hastie and Tibshirani ((1990), page 118). This model can be fitted using `gam()`. For example,

```
> gam(NOx~lf(E,alpha=0.5)+C, data=ethanol)
```

produces a fit that is smooth in E and linear in C.

Using a slightly different motivation, Robinson (1988) and Eubank and Speckman (1993a) arrive at a modified form of (3.3), using $(\mathbf{I} - \mathbf{L}_1)^T (\mathbf{I} - \mathbf{L}_1)$ in place of $\mathbf{I} - \mathbf{L}_1$. Other references on partially linear models include Engle, Granger, Rice and Weiss (1986), Green (1987) and Severini and Staniswalis (1994). A more thorough review is provided by Ichimura and Todd (1999).

3.5.2 Conditionally Parametric Models

The conditionally parametric model is similar to the partially linear model, in that the fit is smooth in some variables and parametric in others. But the conditionally parametric fit allows all coefficients of the parametric variables to depend on the smoothing variables. A conditionally quadratic fit has the form

$$\mu(x, z) = a_0(x) + a_1(x)z + a_2(x)z^2.$$

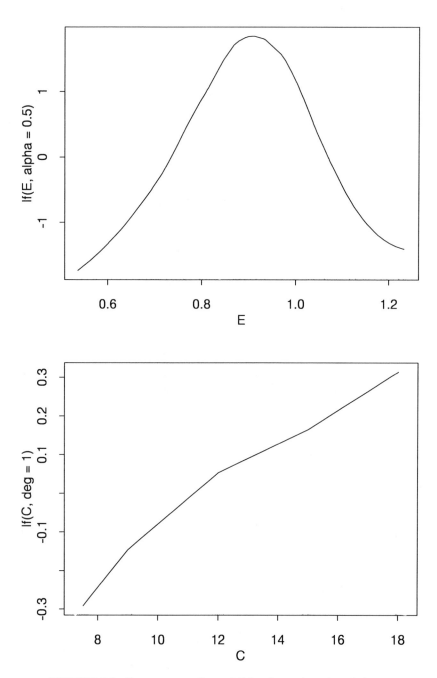

FIGURE 3.3. Components of an additive fit to the ethanol dataset.

For fixed x, $\mu(x, z)$ is a quadratic function of z. But all three coefficients, $a_0(x)$, $a_1(x)$ and $a_2(x)$, are allowed to vary as a function of x. This differs from the partially linear model in which only $a_0(x)$ is allowed to depend on x.

The conditionally parametric fit was considered in detail by Cleveland, Grosse and Shyu (1992) and Cleveland (1994). In particular, they provide a conceptually straightforward way to fit the model: Simply fit a bivariate local regression in x and z but ignore the z variable when computing the distances and smoothing weights. The varying coefficient model (Hastie and Tibshirani 1992) is a broad class of models that encompasses both conditionally parametric fits and partially linear models.

To fit a conditionally parametric model in LOCFIT, one uses the special cpar() function in the model formula. For example,

```
> locfit(NOx~E+cpar(C), data=ethanol, alpha=0.5)
```

produces a fit that is conditionally quadratic in C.

3.6   Exercises

3.1   a) Try to fit the ethanol dataset using local constant and local linear fitting. By varying the bandwidth (using both the fixed and nearest neighbor components, if necessary), can a fit comparable to the local quadratic fit in Figure 2.2 be obtained? Pay attention to both proper modeling of the peak and the leveling off at the boundaries, and to the roughness of the estimates.

   b) Compute the GCV scores for local linear fitting, and compare with the results of local quadratic fitting in Figure 2.7.

3.2   The diabetes dataset used by Hastie and Tibshirani (1990) consists of a predictor variable age and response lcp.

   a) Produce a scatter plot of the data. Does the residual variance look constant?

   b) Fit a local regression model. Construct appropriate smoothed residual plots to investigate the nonhomogeneous variance further. (You'll probably conclude that the nonhomogeneity is real, but much less than might have been guessed from the scatterplot).

   Note: the dataset can be accessed using data=diab in LOCFIT.

3.3   Consider the L_1 cross validation of Exercise 2.8. Write a modified version of gcv() to implement the L_1 generalized cross validation (use residuals() to get $Y_i - \hat{\mu}(x_i)$). Apply this to the NOx~E model for the ethanol dataset, and compare with Figure 2.7.

Hint: See Appendix B.2.

3.4 In this exercise, generalized cross validation is used to compare conditionally parametric fits with bivariate smooth fits for the ethanol dataset.

    a) Make a GCV plot for the model NOx~E+C, with scale=0. Use smoothing parameters ranging from 0.25 to 0.8.

    b) Repeat for the conditionally parametric model NOx~E+cpar(C). Use both the conditionally quadratic (the default) and conditionally linear, by setting deg=1. Compare the results.

    c) scale=0 is equivalent to scale=c(0.204,3.932) (the sample standard deviations). Compute the GCV plot for other scale parameters, such as scale=c(0.204,8). The conditionally parametric fit is obtained as the second component tends to infinity.

3.5 This exercise compares asymptotic and finite sample approximations to the local regression variance.

    a) Generate a sample with $n = 50$, with x_i sampled i.i.d. from the standard normal distribution. Also generate a sample $Y_i \sim N(0, 1)$ (the mean function doesn't matter for this exercise).

    b) Compute a local linear fit, with constant bandwidth $h = 1$. Plot the standard deviation $\|l(x)\|$ using the LOCFIT command plot(fit,what="nlx"). Compute and plot the asymptotic approximation (2.39). Note that

$$\int W(v)^2 dv / (\int W(v) dv)^2 = 175/247$$

for the tricube weight function. Remember the square root!

    c) Repeat using a nearest neighbor bandwidth with $\alpha = 0.7$. When computing the asymptotic variance, approximate the nearest neighbor bandwidth by $h(x) \approx \alpha/(2f(x))$.

    d) Repeat this exercise using two predictor variables, with both components i.i.d. $N(0, 1)$.

4

Local Likelihood Estimation

Generalized linear models (McCullagh and Nelder 1989) provide a generalization of linear regression to likelihood models, for example, when the responses are binary or Poisson count data. Fitting of smooth likelihood models dates to Henderson (1924b), who fitted penalized likelihood models to binary data. This paper, although rarely referred to in modern literature, is particularly noteworthy as it was one of the earliest works on likelihood based regression models.

In this chapter a local likelihood approach is used. This was first proposed in Brillinger (1977) and studied in detail by Tibshirani (1984), Tibshirani and Hastie (1987) and Staniswalis (1989) among others. The local likelihood model is described in Section 4.1. Section 4.2 discusses fitting with LOCFIT. Section 4.3 introduces diagnostic procedures for local likelihood models, including residuals and model assessment criteria. Section 4.4 presents some theoretical results for local likelihood, including existence of the estimates and approximations to the bias and variance.

4.1 The Local Likelihood Model

The likelihood regression model assumes response variables have a density

$$Y_i \sim f(y, \theta_i)$$

where $\theta_i = \theta(x_i)$ is a function of the covariates x_i. Examples include the exponential distribution with mean θ,

$$f(y, \theta) = \frac{1}{\theta} e^{-y/\theta} I_{[0,\infty)}(y)$$

and the discrete Bernoulli distribution with parameter p,

$$f(0, p) = 1 - p; \qquad f(1, p) = p.$$

Let $l(y, \theta) = \log(f(y, \theta))$. The global log-likelihood of a parameter vector $\theta = (\theta(x_1), \dots, \theta(x_n))$ is

$$\mathcal{L}(\theta) = \sum_{i=1}^{n} l(Y_i, \theta(x_i)). \tag{4.1}$$

A generalized linear model assumes $\theta(x)$ has a parametric linear form; for example, $\theta(x) = a_0 + a_1 x$. The local likelihood model no longer assumes a parametric form but fits a polynomial model locally within a smoothing window. The local polynomial log-likelihood is

$$\mathcal{L}_x(a) = \sum_{i=1}^{n} w_i(x) l(Y_i, \langle a, A(x_i - x) \rangle). \tag{4.2}$$

Maximizing over the parameter a leads to the local likelihood estimate.

Definition 4.1 (Local Likelihood Estimate) Let $\hat{a}$ be the maximizer of the local likelihood (4.2). The local likelihood estimate of $\theta(x)$ is

$$\hat{\theta}(x) = \langle \hat{a}, A(0) \rangle = \hat{a}_0.$$

Example 4.1. (Local Logistic Regression.)   Consider the Bernoulli regression model, where

$$P(Y_i = 1) = p(x_i); \qquad P(Y_i = 0) = 1 - p(x_i).$$

The log-likelihood is

$$
\begin{aligned}
l(Y_i, p(x_i)) &= Y_i \log(p(x_i)) + (1 - Y_i) \log(1 - p(x_i)) \\
&= Y_i \log \left(\frac{p(x_i)}{1 - p(x_i)} \right) + \log(1 - p(x_i)).
\end{aligned}
$$

A local polynomial approximation could be used for $p(x_i)$. But this isn't necessarily a good idea, since $0 \le p(x_i) \le 1$, while polynomials have no such constraints. Instead, the interval $(0, 1)$ is mapped to $(-\infty, \infty)$ using the logistic link function

$$\theta(x) = \log \left(\frac{p(x)}{1 - p(x)} \right).$$

Correspondingly, the local polynomial log-likelihood is

$$\mathcal{L}_x(a) = \sum_{i=1}^{n} w_i(x) \left(Y_i \langle a, A(x_i - x) \rangle - \log(1 + e^{\langle a, A(x_i - x) \rangle}) \right).$$

The local polynomial estimate is $\hat{\theta}(x) = \hat{a}_0$. To estimate $p(x)$, the link function is inverted:

$$\hat{p}(x) = \frac{e^{\hat{\theta}(x)}}{1 + e^{\hat{\theta}(x)}}.$$

Definition 4.2 (Link Function) Suppose $f(y, \theta)$ is a parametric family of distributions, with mean

$$\mu = \mu(\theta) = E_\theta(Y).$$

Suppose further that $\mu(\theta)$ is 1-1. The link function is the inverse mapping of this relation; that is, the function $g(\cdot)$ satisfying

$$\theta = g(\mu).$$

The local likelihood estimate of $\mu(x)$ is

$$\hat{\mu}(x) = g^{-1}(\hat{\theta}(x)).$$

In parametric regression models, the choice of link function is largely dictated by the data. If the true mean is log-linear, one has to use the log link. With local regression models, one does not assume the model is globally correct, so the choice of link can be driven more by convenience. One compelling requirement, used to motivate the logistic link in Example 4.1, is that the parameter space for $\theta(x)$ be $(-\infty, \infty)$. For non-negative parameters, the log link is often a natural choice. Another requirement is that $l(y, \theta)$ be concave. This helps ensure stability of the local likelihood algorithm; see Section 4.4.

The variance stabilizing link satisfies

$$-\mathrm{E}\frac{\partial^2}{\partial \theta^2} l(Y, \theta)$$

is constant, independent of the parameter θ. When the link satisfies this property, $\mathrm{var}(\hat{\theta}(x))$ is also independent of $\theta(x)$, at least asymptotically (see Section 4.4). This property is used for confidence interval construction in Section 9.2.3.

Another link, the canonical link, has some attractive theoretical properties. An exponential family of distributions has densities of the form

$$f(y, \mu) = \exp(\tau(\mu)y - \psi(\mu))f_0(y).$$

The canonical link is $\theta = \tau(\mu)$. When a local polynomial model is used for $\theta(x)$, the local likelihood (and hence $\hat{\theta}(x)$) $\mathcal{L}_x(a)$ depends on the data only through $\sum_{i=1}^{n} w_i(x)A(x_i - x)Y_i$. This locally sufficient statistic simplifies theoretical calculations.

4.2   Local Likelihood with LOCFIT

LOCFIT supports local likelihood regression with a variety of families and link functions, as summarized in Table 4.1. By default, a Gaussian family is assumed; this is the standard local regression discussed in Chapter 2.

| | \multicolumn{6}{c}{Link Function} |||||| |
	ident	log	logit	inverse	sqrt	arcsin
Gaussian	d,c,v					
Binomial	y		d,c			v
Poisson	y	d,c			v	
Gamma	y	d,v		c		
Geometric	y	d				
Von Mises	d,v					
Cauchy	d,v					
Huber	d,v					

TABLE 4.1. Supported local likelihood families and link functions: default link (d), canonical link (c), variance stabilizing link (v) and other supported links (y).

Example 4.2. The mine dataset consists of a single response; the number of fractures in the upper seam of coal mines. There are four predictor variables. Fitting log-linear Poisson models, Myers (1990) showed that one predictor variable (percentage of extraction from the lower seam) was highly significant, while two other predictors had some importance. Here, we use the single predictor variable extrp and fit using a local log-linear model. The variable selection problem is considered later.

```
> fit <- locfit(frac~extrp, data=mine, family="poisson",
+   deg=1, alpha=0.6)
> plot(fit, band="g", get.data=T)
```

The Poisson family is specified by the family argument. The default link is the log link (Table 4.1); the plot() method automatically back-transforms to display the estimated mean (Figure 4.1). The plot also shows approximate 95% pointwise confidence intervals for the mean.

The plot shows the mean initially increases, then levels off for extrp > 80. The confidence intervals suggest the leveling off is a real feature; the bands do not cover any curve of the form e^{a+bx}, and thus a log-linear model would appear inadequate for this dataset.

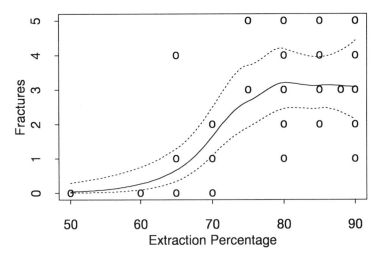

FIGURE 4.1. Mine fracture dataset: local Poisson regression.

Example 4.3.   Mortality data of the type considered in Figure 1.1 is one example of binomial data; the observed mortality rates for each age are the number of deaths divided by the number of patients. Unfortunately, the original source for this dataset did not give the number of patients. Here, we use a second mortality dataset, from Henderson and Sheppard (1919) for which this information is available. The number of trials at each age is given as the weights argument to the locfit() call:

```
> fit <- locfit(deaths~age, weights=n, family="binomial",
+   data=morths, alpha=0.5)
> plot(fit, band="g", get.data = T)
```

Figure 4.2 displays the fit, with 95% pointwise confidence intervals. The data has been smoothed using local quadratic logistic regression, with nearest neighbor span of 0.5. This shows a gradual increasing trend, with some wild behavior at the right boundary. One must be careful when interpreting this plot because there are large differences in the weights. For ages between 70 and 80, there are as many as 150 at-risk patients, but just one for age=99. Likewise, there are just six patients for ages 55 and 56; this (as well as the usual boundary variability) leads to the wide confidence intervals at the left boundary.

We now define the families supported by LOCFIT. Each family is specified using the mean parameter $\mu(x_i)$. Also included is a weight parameter n_i, which for most families can be interpreted as a prior weight or the number of replications for each observation.

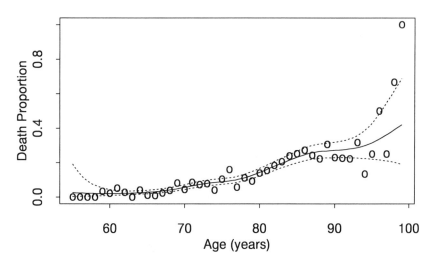

FIGURE 4.2. Local logistic regression for mortality data of Henderson and Sheppard.

The **Gaussian** family has densities

$$f_{Y_i}(y) = \frac{\sqrt{n_i}}{\sqrt{2\pi\sigma^2}} \exp\left(-\frac{n_i}{2\sigma^2}(Y_i - \mu(x_i))^2\right),$$

and the local likelihood criterion is equivalent to the local least squares criterion. Thus, `family="gauss"` produces the local regression estimate, *but assumes* $\sigma^2 = 1$. This distinction is important when constructing confidence intervals; the usual family for local regression is the quasi family `family="qgauss"`. For more discussion of this distinction, see the discussion of quasi-likelihood in Section 4.3.4.

The **binomial** family has probability mass function

$$P(Y_i = y) = \binom{n_i}{y}\mu(x_i)^y(1 - \mu(x_i))^{n_i - y}; y = 0, 1, \dots, n_i. \tag{4.3}$$

The Bernoulli distribution ($n_i = 1$) represents the outcome of a single trial with success probability $\mu(x_i)$. The binomial distribution counts the number of successes in n_i independent trials.

The **Poisson** family is used to model count data. The distribution has the mass function

$$P(Y_i = y) = \frac{(n_i\mu(x_i))^y}{y!}e^{-n_i\mu(x_i)}; y = 0, 1, 2, \dots. \tag{4.4}$$

The **exponential** and **gamma** families (`family="gamma"`) are often used to model survival times. The gamma density function is

$$f_{Y_i}(y) = \frac{\mu(x_i)^{-n_i} y^{n_i - 1}}{\Gamma(n_i)} e^{-y/\mu(x_i)}; y \geq 0. \qquad (4.5)$$

The special case $n_i = 1$ is the exponential distribution.

The **geometric** and **negative binomial** families (`family="geom"`) can be regarded as discrete analogs of the exponential and gamma distributions. The negative binomial distribution has mass function

$$P(Y_i = y) = \binom{n_i + y - 1}{n_i - 1} \frac{\mu(x_i)^y}{(1 + \mu(x_i))^{n_i + y}}; y = 0, 1, \ldots. \qquad (4.6)$$

The geometric distribution is the special case $n_i = 1$. If one observes a sequence of Bernoulli trials with success probability $p(x_i) = \mu(x_i)/(1 + \mu(x_i))$, the geometric distribution models the number of successes observed before a single failure. The negative binomial distribution models the number of successes until n_i failures are observed.

The **von Mises** family (`family="circ"`) has densities

$$f_{Y_i}(y) = \frac{1}{I(n_i)} e^{n_i \cos(y - \mu(x_i))}; -\pi \leq y \leq \pi,$$

where $I(n_i)$ is a normalizing constant. This distribution is frequently used to model datasets where the responses are angular or measured on a circle. Regression models for $\mu(x)$ were introduced by Gould (1969). Fisher (1993) is an extensive resource for statistical methods for circular data.

Numerically, the von Mises family can be difficult to fit, since the log-likelihood has multiple local maxima. If $\hat{\mu}(x)$ is a local likelihood estimate, so is $\hat{\mu}(x) + 2\pi$. More serious problems are caused by adding a linear term. If the x_i are uniform random variables (and hence irrational), some number theoretic arguments show one can come arbitrarily close to interpolation, simply by choosing a linear function with a carefully chosen large slope.

This is related to the barber's pole problem discussed by Gould (1969) and in more detail by Johnson and Wehrly (1978) and Fisher and Lee (1992), who discuss various ways of restricting $\hat{\mu}(x)$ to $[-\pi, \pi]$. None of the solutions seem entirely satisfactory, since $\mu(x)$ may genuinely have multiple circles over the range of the data. For practical purposes, the identifiability problems shouldn't create too much difficulty, unless the data is close to uniform. It also helps if the origin is chosen as a favored direction, so the estimate shouldn't skip from $-\pi$ to π.

The **Cauchy** and **Huber** families are intended mainly for local robust regression. A full description is given in Section 6.4.

4.3   Diagnostics for Local Likelihood

This section discusses diagnostic and model selection issues for local likelihood. Largely, the techniques are natural extensions of the local regression methodology discussed in Section 2.3. Work devoted to diagnostic issues for local likelihood includes Firth, Glosup and Hinkley (1991) and Staniswalis and Severini (1991). The methods are generally similar to techniques used in parametric generalized linear models by McCullagh and Nelder (1989).

4.3.1   Deviance

In Chapter 2, we developed diagnostic methods based on the residuals $Y_i - \hat{\mu}(x_i)$, and the residual sum of squares. For local likelihood models, these tools are less natural. For example, for the gamma family (4.5), $\mu(x)$ is a scale parameter. In this case, it is more natural to consider diagnostics based on the ratio $Y_i/\hat{\mu}(x_i)$ rather than the difference $Y_i - \hat{\mu}(x_i)$.

The natural predictor of a future observation at a point x is $g^{-1}(\hat{\theta}(x))$ where $g(\cdot)$ is the link function. One possible loss function is the **deviance**, for a single observation (x, Y), defined by

$$D(Y, \hat{\theta}(x)) = 2\left(\sup_\theta l(Y, \theta) - l(Y, \hat{\theta}(x))\right).$$

It is easily seen that $D(Y, \hat{\theta}) \geq 0$, and $D(Y, \hat{\theta}) = 0$ if $Y = g^{-1}(\hat{\theta})$. Since it is based on the likelihood, the deviance provides a measure of the evidence an observation Y provides *against* $\hat{\theta}(x)$ being the true value of $\theta(x)$. With a Gaussian likelihood and $\sigma = 1$, the deviance is simply the squared residual.

The total deviance is defined as

$$\sum_{i=1}^n D(Y_i, \hat{\theta}(x_i)). \tag{4.7}$$

This generalizes the residual sum of squares for a regression model.

Example 4.4. Let Y_i be an observation from the gamma family with parameters n_i (known) and μ_i (unknown). The log-likelihood is

$$l(Y_i, \mu_i) = -n_i \log(\mu_i) + (n_i - 1)\log(Y_i) - \frac{Y_i}{\mu_i} - \log(\Gamma(n_i)).$$

For fixed Y_i and n_i, this is maximized at $\mu_i = Y_i/n_i$. Thus, the deviance for an estimate $\hat{\mu}_i$ is

$$D(Y_i, \hat{\mu}_i) = 2\left(-n_i \log(\frac{Y_i}{n_i\hat{\mu}_i}) + \frac{Y_i}{\hat{\mu}_i} - n_i\right).$$

As expected, this depends on Y_i and $\hat{\mu}_i$ only through the ratio $Y_i/\hat{\mu}_i$. Using the Taylor series approximation $\log(x) \approx x - 1 - (x-1)^2/2$ yields

$$D(Y_i, \hat{\mu}_i) \approx \frac{1}{n_i\hat{\mu}_i^2}(Y_i - n_i\hat{\mu}_i)^2.$$

The variance of Y_i is $n_i\mu_i^2$. Thus, the deviance is approximately $(Y_i - E(Y_i))^2/\text{var}(Y_i)$. As $n_i \to \infty$, one has the limiting distribution

$$D(Y_i, \hat{\mu}_i) \Rightarrow \chi_1^2, \tag{4.8}$$

provided $\hat{\mu}_i$ is consistent. This limiting distribution can be generalized to other likelihoods.

4.3.2 Residuals for Local Likelihood

In the case of generalized linear models, a number of suitable extensions of the definition of residuals are discussed in McCullagh and Nelder (1989, section 2.4) and Hastie and Pregibon (1992, page 205). Four possible definitions are:

- Deviance residual

$$r_i = \text{sign}(Y_i - \hat{\mu}_i)D(Y_i, \hat{\theta}_i)^{1/2};$$

- Pearson residual

$$r_i = \frac{Y_i - \hat{\mu}_i}{\sqrt{V_i}};$$

- Response residual

$$r_i = Y_i - \hat{\mu}_i;$$

- Likelihood derivative

$$r_i = \frac{\partial}{\partial\theta}l(Y_i, \hat{\theta}_i),$$

where $\hat{\theta}_i = \hat{\theta}(x_i)$, $\hat{\mu}_i = \hat{\mu}(x_i)$ and $V_i = \text{var}(Y_i)$. For the sample residuals, these are estimated using the fitted values.

For the Gaussian likelihood, all four definitions produce the residuals $Y_i - \mu_i$. For other likelihoods, the definitions do not coincide, and all have slightly different interpretations. The Pearson residuals all have variance 1, and under the assumption $n_i \to \infty$, the residuals are asymptotically $N(0, 1)$. Using (4.8), the deviance residuals have a similar property.

Example 4.5. We compute residuals for the mortality data of Henderson and Sheppard used in Example 4.3. The residuals are found using LOCFIT's `residuals()` function. The type of residual is specified by the type argument; the default is the deviance residuals:

```
> for(ty in c("deviance", "pearson", "response", "ldot")) {
+    res <- residuals(fit, type=ty)
+    plot(morths$age, res, main=ty, type="b")
+    abline(h = 0, lty = 2)
+ }
```

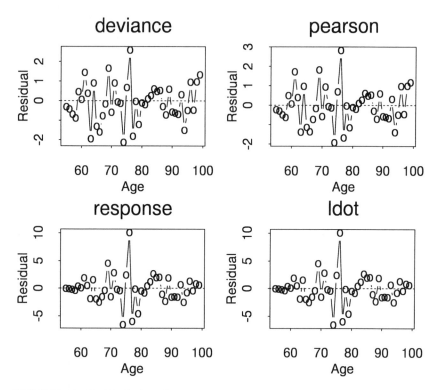

FIGURE 4.3. Residual plots for the mortality data of Henderson and Sheppard.

Figure 4.3 shows four sets of residuals plotted against age. Given the small sample sizes, there is little benefit to smoothing the residual plots, so points are simply joined by lines. No strong patterns appear in the residual plots. Both the deviance and Pearson residuals are mainly in the interval $[-2, 2]$, which indicates that the binomial model adequately models the variability of this dataset.

4.3.3   Cross Validation and AIC

To help guide the choice of local likelihood model, we need extensions of the cross validation and CP methods introduced in Chapter 2. It is natural to consider methods based directly on the likelihood or deviance functions.

Definition 4.3 The **likelihood** (or **deviance**) **cross validation** criterion is defined by substituting the leave-x_i-out estimate $\hat{\theta}_{-i}(x_i)$ in the total deviance (4.7);

$$\text{LCV}(\hat{\theta}) \quad = \quad \sum_{i=1}^{n} D(Y_i, \hat{\theta}_{-i}(x_i))$$

$$= C - 2 \sum_{i=1}^{n} l(Y_i, \hat{\theta}_{-i}(x_i)) \tag{4.9}$$

where C depends on the observations Y_i, but not the estimate $\hat{\theta}(x)$ and hence not the bandwidth or local polynomial degree.

Computation of the n leave-x_i-out estimates can be expensive. An alternative to deletion methods is the method of infinitesimal perturbations, developed in Cook (1977) for linear models, and Pregibon (1981) for logistic regression models. The technique underlies Theorem 2.2, which relates the deletion estimate $\hat{\mu}_{-i}(x_i)$ with the estimate $\hat{\mu}(x_i)$ and the influence function $\text{infl}(x_i)$.

In the local likelihood setting, the simplification of Theorem 2.2 no longer holds. Instead, approximations must be developed; details are provided in Section 4.4.3 and Exercise 4.6. First, we identify an influence function such that

$$\hat{\theta}_{-i}(x_i) \approx \hat{\theta}(x_i) - \text{infl}(x_i)\dot{l}(Y_i, \hat{\theta}(x_i)). \tag{4.10}$$

We use $\dot{l}(y, \theta)$ and $\ddot{l}(y, \theta)$ to denote the first and second partial derivatives of $l(y, \theta)$ with respect to θ. Substituting (4.10) into the deviance and using a one-term Taylor series gives

$$D(Y_i, \hat{\theta}_{-i}(x_i)) \approx D(Y_i, \hat{\theta}_i(x_i)) + 2\text{infl}(x_i)\dot{l}(Y_i, \hat{\theta}(x_i))^2.$$

Summing this over all observations gives an approximation to the likelihood cross validation statistic (4.9). Since $\text{E}(\dot{l}(Y, \theta)^2) = -\text{E}(\ddot{l}(Y, \theta))$, the fitted degrees of freedom are defined as

$$\nu_1 = \sum_{i=1}^{n} \text{infl}(x_i)\text{E}(-\ddot{l}(Y_i, \hat{\theta}(x_i))).$$

This leads to a generalization of the Akaike information criterion (Akaike, 1973, 1974) to local likelihood models.

Definition 4.4 The **Akaike information criterion** (AIC) for local likelihood is

$$\text{AIC}(\hat{\theta}) = \sum_{i=1}^{n} D(Y_i, \hat{\theta}(x_i)) + 2\nu_1 \tag{4.11}$$

where ν_1 is the degrees of freedom for the local likelihood fit.

Example 4.6. We apply the AIC statistic to the mine dataset, sing a variety of nearest neighbor bandwidths:

```
> a <- seq(0.4, 1, by=0.05)
> plot(aicplot(frac~extrp, data=mine, family="poisson",
+    deg=1, alpha=a))
```

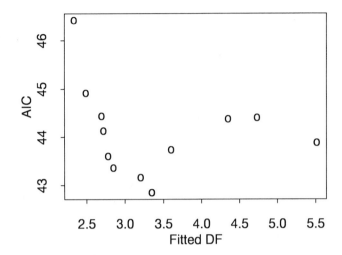

FIGURE 4.4. Akaike information criterion for the mine dataset.

The `aicplot()` function is similar to `gcvplot()` (Section 3.4.2). Figure 4.4 shows the AIC plot. The minimum AIC occurs at about 2.9 degrees of freedom ($\alpha = 0.6$). Larger smoothing parameters (i.e., smaller degrees of freedom) result in inferior fits. This provides evidence that the parametric log-linear model is inadequate for this dataset, and the curvature in Figure 4.1 is real.

4.3.4 Overdispersion

If a likelihood model correctly models a dataset, the Pearson residuals defined in Section 4.3.2 should have mean 0 and variance 1. The deviance residuals are similar, using the approximation of Example 4.4. If the residuals exhibit a nonzero mean (for example, several successive residuals have the same sign), this indicates that the data is oversmoothed, and smaller bandwidths should be used.

Overdispersion occurs when the residuals have variance larger than 1. For example, the Poisson distribution has the property $\text{var}(Y_i) = \text{E}(Y_i)$. But count data often exhibit more variability than this relation can explain. The mean can still be estimated using Poisson regression, but the variance of $\hat{\mu}(x)$ may be severely underestimated.

There are several ways to handle overdispersed data. One method is through a variance stabilizing transformation, where one finds a function $g(Y)$ such that the transformed data $g(Y_i)$ has approximately constant variance. A local regression model is then fitted to the transformed data. The most commonly used family of transformations is the Box-Cox, or power, family (Box and Cox 1964). A more sophisticated implementation is the ATS (average, transformation and smoothing) method of Cleveland, Mallows and McRae (1993), which includes a presmoothing step prior to the transformation.

Another technique is to find a family of distributions that better fits the data. For example, the negative binomial distribution (4.6) has mean $w\mu$ and variance $w\mu(1 + \mu)$; in this case, the variance is always larger than the mean. One then estimates the shape parameter w and fits the corresponding negative binomial model. An example using this approach is provided in Section 7.3.1.

A cleaner solution is quasi-likelihood, introduced by Wedderburn (1974); see also chapter 9 of Wedderburn (1974) and the recent book by Heyde (1997). Fan, Heckman and Wand (1995) discuss the local quasi-likelihood method. In quasi-likelihood models, one assumes a relation between the mean and variance of the observations:

$$\text{var}(Y_i) = \sigma^2 V(\mu_i)$$

where $V(\mu)$ is a known function and σ^2 is an unknown dispersion parameter. For example, under a Poisson model, one has $\text{var}(Y_i) = \mu_i$, so the quasi-Poisson model takes $V(\mu) = \mu$. Table 4.2 summarizes the variance relationships for the common families supported in LOCFIT. In locfit() calls, the quasi-family is obtained, for example, with the family="qpoisson" argument.

Family	Variance $\sigma^2 V(\mu)$
quasi-Gaussian	σ^2
quasi-binomial	$\sigma^2\mu(1 - \mu)$
quasi-Poisson	$\sigma^2\mu$
quasi-gamma	$\sigma^2\mu^2$
quasi-geometric	$\sigma^2\mu(\mu + 1)$

TABLE 4.2. Quasi-likelihood families and their variance functions.

Note that fitting a quasi-likelihood model is identical to fitting the corresponding likelihood model. The difference is in variance estimation: While the likelihood families assume the dispersion parameter is $\sigma^2 = 1$, the quasi-likelihood families estimate the dispersion parameter. The estimate used by LOCFIT is

$$\hat{\sigma}^2 = \frac{n}{n - 2\nu_1 + \nu_2} \frac{\sum_{i=1}^{n} \dot{l}(Y_i, \hat{\theta}(x_i))^2}{\sum_{i=1}^{n} \ddot{l}(Y_i, \hat{\theta}(x_i))}.$$

4.4 Theory for Local Likelihood Estimation

This section addresses some of the theoretical issues concerning local likelihood. Our emphasis is on results that have immediate practical consequences. First, we look at the motivation for maximizing the local likelihood. Then, we turn to important computational concerns and related issues such as existence and uniqueness. Finally, approximate representations for the estimate are derived; this leads to bias and variance approximations, and definitions of degrees of freedom.

4.4.1 Why Maximize the Local Likelihood?

The log-likelihood $\mathcal{L}(\theta)$, for fixed θ, is a random variable, dependent on the observations $Y_1, \ldots, Y_n$. The mean $E(\mathcal{L}(\theta))$ is a function of the parameter vector θ, and this mean function is maximized at the true parameter vector θ. For any parameter vector θ^*, Exercise 4.2 shows that

$$E\left(\mathcal{L}(\theta^*)\right) \leq E\left(\mathcal{L}(\theta)\right). \tag{4.12}$$

This motivates maximum likelihood: parameter values θ for which $\mathcal{L}(\theta)$ are the most likely values of θ, given the observed data. Thus, among a class of candidate parameter vectors, we select the one that maximizes the empirical log-likelihood.

This maximum likelihood property extends to the local log-likelihood:

$$E\left(\sum_{i=1}^{n} w_i(x)l(Y_i, \theta_i^*)\right) \leq E\left(\sum_{i=1}^{n} w_i(x)l(Y_i, \theta_i)\right) \tag{4.13}$$

with equality if and only if $\theta_i^* = \theta_i$ for all i with $w_i(x) > 0$. The local likelihood estimate considers candidate classes of the form $\theta_i^* = \langle a, A(x_i - x) \rangle$ and maximizes over this class of candidates.

4.4.2 Local Likelihood Equations

Assuming the likelihood is nicely behaved, the parameter vector $\hat{a}$ is a solution of the local likelihood equations

$$\sum_{i=1}^{n} w_i(x)A(x_i - x)\dot{l}(Y_i, \langle a, A(x_i - x) \rangle) = 0, \tag{4.14}$$

obtained by differentiating (4.2). In matrix notation, the local likelihood equations can be written

$$\mathbf{X}^T \mathbf{W} \dot{l}(Y, \mathbf{X}a) = 0 \tag{4.15}$$

where, as before, $\mathbf{X}$ is the design matrix and $\mathbf{W}$ is the diagonal matrix with entries $w_i(x)$.

For most likelihoods, the local likelihood equations (4.14) do not have a closed form solution, and must be solved by iterative methods. This leads to two questions:

1. Does the maximizer $\hat{a}$ exist?

2. Is the maximizer $\hat{a}$ unique?

The following theorem addresses these questions for concave likelihoods.

Theorem 4.1 Suppose the log-likelihood $l(y, \theta)$ is defined for θ in an open interval (a, b) ($a = -\infty$ and $b = \infty$ are permitted); $l(y, \theta)$ has a continuous derivative with respect to θ and $l(y, \theta) \to -\infty$ as $\theta \downarrow a$ or $\theta \uparrow b$. Further, suppose $\mathbf{WX}$ has full column rank. Then the maximizer $\hat{a}$ exists and satisfies the local likelihood equations (4.14). If in addition $l(y, \theta)$ is concave, the solution of (4.14) is unique.

Proof: Let $a^{(j)}$ be a sequence of parameter estimates such that

$$\lim_{j \to \infty} \mathcal{L}_x(a^{(j)}) = \sup_a \mathcal{L}_x(a). \tag{4.16}$$

If $a^{(j)}$ has a limit point a^*, then by continuity, $\mathcal{L}_x(a^*) = \sup_a \mathcal{L}_x(a)$; hence $a^* = \hat{a}$. Otherwise, $\|a^{(j)}\| \to \infty$; since $\mathbf{WX}$ has full rank, this implies $\theta_i^{(j)} = \langle a^{(j)}, A(x_i - x) \rangle \to \pm\infty$ for some i with $w_i(x) > 0$. But since $l(y, \theta)$ is bounded above, this contradicts (4.16).

Since the parameter space is open, $\hat{a}$ lies in the interior, and thus is a solution of the local likelihood equations. Differentiating (4.15) yields the Jacobian matrix $-\mathbf{J}_1(\mathbf{X}a)$, where

$$\begin{aligned}
\mathbf{J}_j(\theta) &= -\sum_{i=1}^{n} W\left(\frac{x_i - x}{h}\right)^j A(x_i - x) A(x_i - x)^T \ddot{l}(Y_i, \theta_i) \\
&= \mathbf{X}^T \mathbf{W}^j \mathbf{V} \mathbf{X} \tag{4.17}
\end{aligned}$$

and $\mathbf{V}$ is a diagonal matrix with elements $-(\ddot{l}(Y_i, \theta_i))$. The concavity of $\ddot{l}(Y_i, \theta_i)$ implies $\mathbf{J}_1(\theta)$ is positive definite; strictly so since $\mathbf{X}^T\mathbf{W}$ has full rank. This implies uniqueness of $\hat{a}$. $\square$

Theorem 4.1 gives a number of conditions on the choice of parameterization that help ensure the local likelihood estimation is well behaved. Unfortunately the conditions are rather restrictive; particularly for discrete families. Fortunately, modifying the results for specific families is usually straightforward. Exercises 4.3 and 4.4 study the Poisson and Bernoulli families more closely.

4.4.3 Bias, Variance and Influence

Because of the nonlinear definition of $\hat{a}$, it is not possible to derive exact means and variances of $\hat{a}$; indeed, these often don't exist because of singularities that occur with small probabilities. For example, in the binomial family, there is always a positive probability that all responses are 0, in which case the local likelihood estimate does not exist. We still need distributional approximations for the local likelihood estimate, and to make headway we need approximations to the estimate itself. We should emphasize the approximations derived here depend on the *design points* $x_1, \ldots, x_n$, and not on an asymptotic design density. This is quite different from previous results in Fan, Heckman and Wand (1995) and Fan and Gijbels (1996, pages 196-197).

The results proceed in three parts. First, Theorem 4.2 establishes consistency of the local likelihood estimate. Theorem 4.3 establishes the asymptotic representation of the estimate, from which variance approximations can be derived. Theorem 4.4 derives a bias approximation using derivatives of $\theta(x)$.

Theorem 4.2 Suppose $l(y, \theta)$ is concave, bounded and twice differentiable for all y. Then for either random or regular designs,

$$\mathbf{H}\hat{a} \xrightarrow{p} \begin{pmatrix} \theta(x) \\ 0 \\ \vdots \\ 0 \end{pmatrix}$$

aas $h \to 0$ and $nh^d \to \infty$. Here, $\mathbf{H}$ is a diagonal matrix of powers of h; $\mathbf{H}A(v/h) = A(v)$.

Remark: This result implies consistency of $\hat{\theta}(x) = \hat{a}_0$. It does not imply the remaining elements of $\hat{a}$ converge to 0.

Proof: Applying the weak law of large numbers and using the continuity of $\theta(x)$ one obtains, for any fixed vector a,

$$\frac{1}{nh^d} \sum_{i=1}^{n} W\left(\frac{x_i - x}{h}\right) l(Y_i, \left\langle a, A(\frac{x_i - x}{h})\right\rangle)$$

$$\xrightarrow{p} \frac{f(x)}{h^d} \int \int W\left(\frac{u - x}{h}\right) l(y, \left\langle a, A(\frac{u - x}{h})\right\rangle) e^{l(y,\theta(x))} dy du$$

$$= \int \int W(v) l(y, \langle a, A(v)\rangle) e^{l(y,\theta(x))} dy dv$$

where $f(x)$ is the design density. The left-hand side is maximized at $a = \mathbf{H}\hat{a}$, while an argument similar to (4.13) shows the right-hand side is maximized at $(\theta(x), 0, \ldots, 0)^T$. The theorem follows using convexity of the likelihood. $\qquad\square$

The components of the vector $\hat{a}$ should estimate the coefficient vector $\tilde{a}$ of the Taylor series expansion of $\theta(\cdot)$ expanded around the fitting point x. As a first step in obtaining an asymptotic representation, we look at the discrepancy $\hat{a} - \tilde{a}$. This leads to the following result.

Theorem 4.3 Under the conditions of Theorem 4.2,

$$\mathbf{H}(\hat{a} - \tilde{a}) = \mathbf{H}\mathbf{J}_1^{-1}\mathbf{X}^T\mathbf{W}\dot{l}(Y, \mathbf{X}\tilde{a}) + o((nh^d)^{-1/2}). \qquad (4.18)$$

Proof: Expanding the local likelihood equations yields

$$\begin{aligned} 0 &= \mathbf{H}^{-1}\mathbf{X}^T\mathbf{W}\dot{l}(Y, \mathbf{X}\hat{a}) \\ &= \mathbf{H}^{-1}\mathbf{X}^T\mathbf{W}\dot{l}(Y, \mathbf{X}\tilde{a}) - \mathbf{H}^{-1}\mathbf{J}_1(\hat{a} - \tilde{a}) + o(nh^d\mathbf{H}(\hat{a} - \tilde{a})), \end{aligned}$$

and hence

$$\mathbf{H}(\hat{a} - \tilde{a}) = \mathbf{H}\mathbf{J}_1^{-1}\mathbf{X}^T\mathbf{W}\dot{l}(Y, \mathbf{X}\tilde{a}) + o(\mathbf{H}(\hat{a} - \tilde{a})).$$

The result follows since $\mathbf{H}\mathbf{J}_1^{-1}\mathbf{X}^T\mathbf{W}\dot{l}(Y, \mathbf{X}\tilde{a})$ has size $O_p((nh^d)^{-1/2})$.   □

In Theorem 4.3, the first row of the matrix $\mathbf{J}_1^{-1}\mathbf{X}^T\mathbf{W}$ plays a role similar to the weight diagram in local regression. The influence function at x_i is defined to be the ith component of this weight diagram:

$$\mathrm{infl}(x) = W(0)e_1^T\mathbf{J}_1^{-1}e_1.$$

This measures the sensitivity of the estimate $\hat{\theta}(x_i)$ to changes in $\dot{l}(Y_i, \theta_i)$. A rather more subtle interpretation of the influence function is the leave-x_i-out cross validation approximation

$$\hat{\theta}_{-i}(x_i) = \hat{\theta}(x_i) - \mathrm{infl}(x_i)\dot{l}(Y_i, \hat{\theta}_i);$$

see Exercise 4.6. One also obtains an approximate variance of $\hat{\theta}(x_i)$ from Theorem 4.3:

$$\mathrm{vari}(x) = e_1^T\mathbf{J}_1^{-1}\mathbf{J}_2\mathbf{J}_1^{-1}e_1. \qquad (4.19)$$

The fitted degrees of freedom for a local likelihood model are defined as

$$\begin{aligned} \nu_1 &= \sum_{i=1}^n \mathrm{infl}(x_i)v_i \\ \nu_2 &= \sum_{i=1}^n \mathrm{vari}(x_i)v_i \end{aligned} \qquad (4.20)$$

where $v_i = -\ddot{l}(Y_i, \theta(x_i))$. One may prefer to use $\mathrm{E}(v_i)$ in place of v_i in (4.20) and the matrices $\mathbf{J}_j$, since the expected values are nonrandom and necessarily positive, even when the log-likelihood is not concave. This is essentially the question of observed versus expected Fisher information in parametric models, and makes little difference asymptotically.

The final step in the asymptotic representation is to identify the bias of the local likelihood estimate. This can be expressed using higher order derivatives of $\theta(x)$. The result is stated for one dimensional x_i; the multivariate result involves terms for all partial derivatives.

Theorem 4.4 The first term of the bias expansion is

$$\mathrm{E}(\mathbf{HJ}_1^{-1}\mathbf{X}^T\mathbf{W}\dot{l}(Y,\mathbf{X}\tilde{a}))$$
$$= \frac{\theta^{(p+1)}(x)}{(p+1)!}\mathbf{HJ}_1^{-1}\sum_{i=1}^{n}w_i(x)(x_i-x)^{p+1}A(x_i-x)v_i + o(h^{p+1}).$$

For $p \geq 1$, the second term involving $\theta^{(p+2)}$ is similar.

Proof: Let $\tilde{\theta}_i = \langle \tilde{a}, A(x_i-x)\rangle$. Then

$$\theta(x_i) = \tilde{\theta}_i + \frac{(x_i-x)^{p+1}}{(p+1)!}\theta^{(p+1)}(x) + O(h^{p+2})$$

uniformly on the smoothing window, and

$$\begin{aligned}
\dot{l}(Y_i,\tilde{\theta}_i) &= \dot{l}(Y_i,\theta(x_i)) + (\tilde{\theta}_i - \theta(x_i))\ddot{l}(Y_i,\theta(x_i)) + O((\theta(x_i)-\tilde{\theta}_i)^2)\\
&= \dot{l}(Y_i,\theta(x_i)) - \frac{(x_i-x)^{p+1}}{(p+1)!}\theta^{(p+1)}(x)\ddot{l}(Y_i,\theta(x_i)) + O(h^{p+2}).
\end{aligned}$$

Substituting into Theorem 4.3 and remembering $\mathrm{E}(\dot{l}(Y_i,\theta(x_i))) = 0$ completes the proof. □

We remark that the careful theoretical analysis of local likelihood is important. Many statistical software packages include functions for fitting generalized linear models: the `glm()` function in S-Plus, and similar functions in other packages. These functions usually allow weights for each observation, so local likelihood models can be fitted by calling `glm()` repeatedly, with a new set of weights for each fitting point. This implementation was used by Bowman and Azzalini (1997) and the associated software.

This approach produces correct estimates but incorrect inferences. The problem is that `glm()` interprets weights as a sample size; for example, the n_i in (4.3). This appears as a multiplier for the $\mathbf{V}$ matrix in the Jacobian (4.17), rather than as the required $\mathbf{W}$. In particular, this implies the matrix $\mathbf{J}_2$ is computed incorrectly, and the standard errors are not correct, even asymptotically.

4.5   Exercises

4.1 This exercise uses the Henderson and Shepherd mortality dataset, from Example 4.3.

a) Compute a local quadratic fit, using the arcsin link. Plot the fit and confidence intervals. Compare with Figure 4.2. Explain the narrower confidence intervals near the left boundary.

b) Compute and compare AIC and LCV plots for both the logistic and arcsin links. Use both local quadratic and local linear fitting. Which fits appear best? Does a global linear model (with either link function) appear satisfactory?

4.2 a) Prove for any a, b,

$$\log(a) \leq \log(b) + \frac{a-b}{b}. \qquad (4.21)$$

b) Suppose a random variable Y has density $g(y)$, and let $g^*(y)$ be any other density. Show that

$$E(\log g^*(Y)) \leq E(\log g(Y))$$

with equality if and only if $g = g^*$ almost everywhere.

c) Prove (4.12) and (4.13).

4.3 For the Poisson family, the conditions of Theorem 4.1 are not satisfied when $Y_i = 0$ for some i, since $l(0, \mu) = -\mu$ is monotone.

a) Using the canonical link $\theta = \log(\mu)$, show the result of Theorem 4.1 still holds, with the additional requirement that $\mathbf{W}\mathbf{X}$ have full rank after deleting rows corresponding to $Y_i = 0$.

b) Show the existence extends to the identity and square root links. Provide an example to show the estimate might not satisfy the local likelihood equations.

4.4 For the Bernoulli family, the situation is even worse, since the likelihood is monotone for all observations. Using local linear fitting with the logistic link, show the local likelihood estimate exists if and only if no $\gamma \neq 0$ and c exists for which

$$\langle \gamma, x_i \rangle \leq c \quad \forall \quad i \text{ with } w_i(x) > 0, Y_i = 0$$
$$\langle \gamma, x_i \rangle \geq c \quad \forall \quad i \text{ with } w_i(x) > 0, Y_i = 1;$$

that is, no hyperplane separates the observations with $Y_i = 0$ from those with $Y_i = 1$.

4.5 Consider Bernoulli trials (x_i, Y_i) with $Y_i \in \{0, 1\}$ and replicated x values. The dataset can be smoothed directly using logistic regression or replicated x values pooled to form a new dataset (x_j^*, n_j, Y_j^*) using n_j as the weights argument.

a) If the same bandwidths are used for each dataset, show the same estimate results. Also show the influence function is the same for each dataset.

b) Show the likelihood cross validation scores for the two datasets are unequal, so that minimizing $\mathrm{LCV}(\hat{\theta})$ may yield two different answers. Show $\mathrm{AIC}(\hat{\theta})$ is the same, up to an additive constant (independent of $\hat{\theta}$).

4.6 This exercise develops the method of infinitesimal perturbations and derives the approximation (4.10). Consider the local likelihood estimate at a point $x = x_i$ and the modified local likelihood equations

$$\mathbf{X}^T \mathbf{W} \dot{l}(Y, \mathbf{X}a) - \lambda W(0)e_1 \dot{l}(Y_i, \langle a, A(0) \rangle) = 0$$

where λ is a parameter and the solution is $\hat{a}(\lambda)$.

a) Show $\hat{a}(0)$ is the full local likelihood parameter estimate, while $\hat{a}(1)$ is the leave-x_i-out parameter estimate.

b) Show

$$\left. \frac{\partial \hat{a}(\lambda)}{\partial \lambda} \right|_{\lambda=0} = \mathbf{J}^{-1} e_1 W(0) \dot{l}(Y_i, \hat{\theta}(x_i)).$$

c) Conclude, to a first order approximation, that

$$\hat{\theta}_{-i}(x_i) \approx \hat{\theta}(x_i) - \mathrm{infl}(x_i) \dot{l}(Y_i, \hat{\theta}(x_i)),$$

and hence

$$\mathrm{LCV}(\hat{\theta}) \approx \sum_{i=1}^{n} D(Y_i, \hat{\theta}(x_i)) + 2 \sum_{i=1}^{n} \mathrm{infl}(x_i) \dot{l}(Y_i, \hat{\theta}(x_i))^2.$$

5
Density Estimation

Suppose observations $X_1, \ldots, X_n$ have an unknown density $f(x)$. The density estimation problem is to estimate $f(x)$.

The histogram is a density estimate, where the x space is divided into bins, and counts of the data are provided for each bin. This is a simple and intuitive approach, but it has problems for continuous data. How do we choose the bins, and where should they be placed? A discrete histogram may smooth out important features in the data.

This chapter studies an adaptation of the local likelihood method to density estimation. Section 5.1 derives the estimate. Section 5.2 describes the implementation, using the LOCFIT software. Section 5.3 introduces diagnostic methods such as residual plots and AIC. The more technical Section 5.4 studies theoretical properties for the local likelihood estimate.

5.1   Local Likelihood Density Estimation

An extension of local likelihood methods to the density estimation problem is described in Loader (1996b) and Hjort and Jones (1996). Consider the log-likelihood function

$$\mathcal{L}(f) = \sum_{i=1}^{n} \log(f(X_i)) - n\left(\int_{\mathcal{X}} f(u)du - 1\right) \tag{5.1}$$

where $\mathcal{X}$ is the domain of the density. The definition (5.1) of the log-likelihood is unusual, with the added a penalty term $n(\int_{\mathcal{X}} f(u)du - 1)$. If f

is a density, the penalty is 0, so (5.1) coincides with the usual log-likelihood in this case. The reason for adding the penalty to (5.1) is that $\mathcal{L}(f)$ can be treated as a likelihood for any non-negative function f without imposing the constraint $\int f(x)dx = 1$. A more complete justification is given in Section 5.4.

A localized version of the log-likelihood is

$$\mathcal{L}_x(f) = \sum_{j=1}^{n} W\left(\frac{X_j - x}{h}\right) \log(f(X_j)) - n \int_{\mathcal{X}} W\left(\frac{u - x}{h}\right) f(u)du. \quad (5.2)$$

We consider a local polynomial approximation for $\log(f(u))$; $\log(f(u)) \approx \langle a, A(u - x)\rangle$ in a neighborhood of x. The local likelihood becomes

$$\mathcal{L}_x(a) = \sum_{j=1}^{n} W\left(\frac{X_j - x}{h}\right) \langle a, A(X_j - x)\rangle$$

$$-n \int_{\mathcal{X}} W\left(\frac{u - x}{h}\right) \exp(\langle a, A(u - x)\rangle)du. \quad (5.3)$$

Definition 5.1 Let $\hat{a} = (\hat{a}_0, \ldots, \hat{a}_p)^T$ be the maximizer of the local log-likelihood (5.3). The **local likelihood density estimate** is defined as

$$\hat{f}(x) = \exp(\langle \hat{a}, A(0)\rangle) = \exp(\hat{a}_0). \quad (5.4)$$

Under fairly general conditions, the local parameter vector $\hat{a}$ is the solution of the system of local likelihood equations obtained by differentiating (5.3):

$$\frac{1}{n}\sum_{j=1}^{n} A(X_j - x)w_j(x)$$

$$= \int_{\mathcal{X}} A(u - x)W\left(\frac{u - x}{h}\right) e^{\langle \hat{a}, A(u-x)\rangle}du \quad (5.5)$$

where $w_j(x) = W((X_j - x)/h)$. These equations have a simple and intuitive interpretation. The left-hand side of (5.5) is simply a vector of localized sample moments up to order p, while the right-hand side is localized population moments using the log-polynomial density approximation. The local likelihood estimate simply matches localized sample moments with localized population moments.

Example 5.1. (Local Constant Fitting). When the local constant polynomial ($p = 0$) is used, (5.5) consists of the single equation

$$\frac{1}{n}\sum_{j=1}^{n} w_j(x) = \int_{\mathcal{X}} W\left(\frac{u - x}{h}\right) \exp(\hat{a}_0)du,$$

yielding the closed form for the density estimate

$$\hat{f}(x) = \exp(\hat{a}_0) = \frac{1}{nh \int W(v)dv} \sum_{j=1}^{n} w_j(x). \tag{5.6}$$

This is the kernel density estimate considered by Rosenblatt (1956), Whittle (1958) and Parzen (1962).

The kernel density estimate has been widely studied; see, for example, the books by Prakasa Rao (1983), Silverman (1986), Scott (1992) and Wand and Jones (1995). Being based on a local constant approximation, it suffers from the same problems as local constant regression, such as trimming of peaks. An additional problem occurs in the tails, since increasing bandwidths for data sparsity can lead to severe bias. This problem was investigated more fully by Loader (1996b), where relative efficiencies of kernel and local log-polynomial methods were compared.

5.1.1   Higher Order Kernels

The system of equations (5.5) defining the local likelihood estimate has the simple moment-matching interpretation noted previously. The moment matching equations can also be used with other local approximations to the density. The identity link $f(u) \approx \langle a, A(u - x) \rangle$ gives the system

$$\frac{1}{n} \sum_{j=1}^{n} A(X_j - x)w_j(x) = \int_{\mathcal{X}} A(u - x)W\left(\frac{u - x}{h}\right) \langle \hat{a}, A(u - x) \rangle \, du, \tag{5.7}$$

with the density estimate being $\hat{f}(x) = \hat{a}_0$. Since (5.7) is a linear system of equations, one can solve explicitly for $\hat{a}$ and $\hat{f}(x)$. Local approximation estimates of this type were considered in Sergeev (1979).

Some manipulation shows the solution of (5.7) can be written

$$\hat{f}(x) = \frac{1}{nh} \sum_{i=1}^{n} W^*\left(\frac{X_i - x}{h}\right)$$

where $W^*(v) = \langle \beta, A(v) \rangle W(v)$ for an appropriate coefficient vector β. The kernel $W^*(v)$ satisfies the moment conditions,

$$\int W^*(v)dv = 1$$

$$\int v^j W^*(v)dv = 0, j = 1, \ldots, p. \tag{5.8}$$

Weight functions satisfying these moment conditions are known as kernels of order $p + 1$, and were introduced by Parzen (1962). The motivation is bias reduction: If the bias of $\hat{f}(x)$ is expanded in a Taylor series, the moment conditions (5.8) ensure that the low order terms are zero. The close connection between density estimation using higher order kernels and local polynomial fitting was investigated by Lejeune and Sarda (1992).

For practical purposes, the higher order kernel estimates tend to be less satisfactory than the local likelihood approach based on (5.5). The reason is that (5.5) applies a local polynomial approximation for $\log(f(x))$ rather than $f(x)$ itself. Since $f(x)$ must be non-negative, the polynomial approximation for $\log(f(x))$ is usually better, particularly in the tails of densities.

5.1.2 Poisson Process Rate Estimation

A problem closely related to density estimation is estimating the intensity function for a point process. If $X_1, \ldots, X_N$ are the random points of a point process, the corresponding counting process is

$$Z(A) = \sum_{i=1}^{N} I(X_i \in A)$$

for any set A. The intensity function, $\lambda(x)$, defines the mean of $Z(A)$:

$$E(Z(A)) = \int_A \lambda(x)dx. \tag{5.9}$$

A simple example of a point process is the nonhomogeneous Poisson process, where $Z(A)$ has a Poisson distribution with mean (5.9). For this process, the log-likelihood function is

$$\mathcal{L}(\lambda, N) = \sum_{i=1}^{N} \log \lambda(X_i) - \int_{\mathcal{X}} \lambda(x)dx.$$

See, for example, Cox and Lewis (1966). This differs from the likelihood (5.1) for density estimation in only one important respect: the dropping of the factor N in front of the integral. The localization of the likelihood and derivation of the local likelihood equations follow similarly, and the implementation of the estimation procedure is almost identical.

5.1.3 Discrete Data

In practice, all datasets are discrete. For the types of measurements usually modeled as coming from continuous distributions, this discreteness is often at a very fine level and can be ignored. With more heavily rounded data, the discreteness becomes important, and it must be modeled using a discrete probability mass function rather than a continuous density. Smooth probability estimates of a mass function have been widely studied using kernel methods; see, for example, Dickey (1968), Aitchison and Aitken (1976), Titterington (1980) and Simonoff (1987, 1995, 1996). The last of these also considers local likelihood approaches.

A local log-likelihood for the mass function is obtained by replacing the integrals in (5.1) and (5.2) by sums over the mass points. Assume the data points $X_1, \ldots, X_n$ are integer valued, and consider the (j, Y_j) pairs, where Y_j is the number of observations equal to j. The total number of observations is $n = \sum_{-\infty}^{\infty} Y_j$. The corresponding probabilities to be estimated are $p(j) = P(X_1 = j)$. Using a local polynomial model for $\log(p(j))$ in a neighborhood of a fitting point x, the discrete version of the local likelihood (5.2) is

$$\mathcal{L}_x(a) = \sum_{j=-\infty}^{\infty} W\left(\frac{j-x}{h}\right) \langle a, A(j-x) \rangle Y_j \tag{5.10}$$

$$-n \sum_{j=-\infty}^{\infty} W\left(\frac{j-x}{h}\right) e^{\langle a, A(j-x) \rangle}. \tag{5.11}$$

This is the local likelihood (4.2), with $l(y, \mu) = y \log(\mu) - n\mu$. Except for the factor n, this is the Poisson log-likelihood. Thus, estimating a mass function is almost equivalent to a local Poisson regression. Note the sum on the right-hand side of (5.11) is *not* restricted to values of j with $Y_j > 0$.

Although the close relation between discrete probability estimation, Poisson regression and density estimation is apparent, there are important theoretical differences. The raw probability Y_j/n is a $\sqrt{n}$-consistent estimate of $p(j)$, and, *given a sufficiently large sample*, this will be the best local estimate. Thus, the large sample behavior of the continuous density and discrete probability estimates are quite different.

Discreteness also has a major impact on bandwidth selection. This topic will be discussed more later, but the important point is that *discrete distributions do not have densities*. Thus, if a selector designed for continuous data is blindly applied to discrete data, problems *should* result, as the selector will prefer densities that place a spike at each data point. Selectors have to be adapted specifically to discrete data, and the result $h = 0$ (that is, use the raw probabilities) has to be considered a legitimate answer.

5.2 Density Estimation in LOCFIT

In LOCFIT, density estimation corresponds to `family="density"`. This family becomes the default when no left-hand side is specified in the model formula. Using `family="rate"` gives the Poisson process rate estimate.

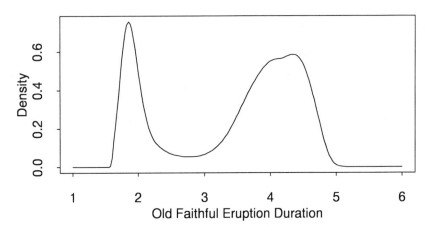

FIGURE 5.1. Density estimation for the Old Faithful geyser dataset.

Example 5.2. The Old Faithful geyser dataset, as given by Weisberg (1985) and Scott (1992), contains the durations of 107 eruptions. The density is estimated using a mixed smoothing parameter with a fixed component of 0.8 and nearest neighbor span of 0.1:

```
> fit <- locfit(~geyser,alpha=c(0.1,0.80),flim=c(1,6))
> plot(fit, mpv=200, xlab="Old Faithful Eruption Duration",
+   ylab="Density", get.data=T)
```

The fit is shown in Figure 5.1. This clearly shows two peaks in the data: a sharp peak around two minutes and a broader peak around 4 minutes. Note the `flim=c(1,6)` argument given to the `locfit()` call; this specifies fitting limits slightly outside the range of the data, thus allowing us to see the tails of the density. The `get.data=T` option causes the data points to be displayed as a 'rug' along the bottom of the plot, rather than the scatter plot used in the regression setting.

Example 5.3. The high order kernels discussed in Section 5.1.1 can be fitted using `link="ident"`. We use the fourth order kernel (local quadratic) estimate for the Old Faithful dataset:

```
> fit <- locfit(~geyser, alpha=c(0.1,0.6), flim=c(1,6),
+   link="ident")
> plot(fit, mpv=200, xlab="Old Faithful Eruption Duration",
+   ylab="Density", get.data=T)
```

The resulting fit in Figure 5.2 seems less satisfactory than that obtained previously in Figure 5.1: The estimate is not constrained to be positive,

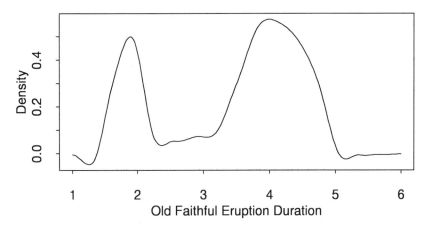

FIGURE 5.2. Local quadratic (fourth order kernel) fit to the Old Faithful geyser dataset.

and the method seems to oversmooth the left peak, despite the use of a smaller bandwidth.

Example 5.4. Izenman and Sommer (1988) and Sheather (1992) report a dataset on measurements of the thickness of 486 postage stamps of the 1872 Hidalgo issue of Mexico. The thicknesses are recorded to the nearest 0.001 millemetres. This discreteness is coarse enough to matter, as is seen when bandwidth selectors are applied to this problem (Exercise 5.4). A local qudratic density estimate is computed using the Poisson regression model:

```
> fit <- locfit(count~thick, weights=rep(0.486,76),
+    data=stamp, family = "poisson", alpha = c(0, 0.004))
> plot(fit, m=200, get.data=T)
```

The critical point is the `weights` argument. Setting `weights=rep(n,76)` effectively divides the Poisson regression by n, leading to estimation of the mass function. The probability of a bin centered at a point x_i is approximately $n\Delta f(x_i)$ where Δ is the size of the bin and $f(x)$ is the density. Comparing with the Poisson family (4.4), we set the weight $n_i = n\Delta$, and the mean $\mu(x_i) = f(x_i)$. In this example, $n = 486$ and $\Delta = 0.001$.

Figure 5.3 shows the resulting multimodal estimate. The explanation for the multimodality, provided by Izenman and Sommer (1988), is that a large number of different types of paper were used to print this stamp.

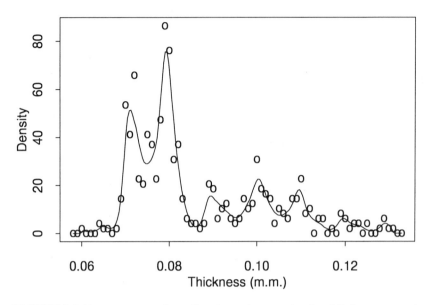

FIGURE 5.3. Postage stamp data. Density estimate using local Poisson regression for discrete data.

5.2.1 Multivariate Density Examples

Multivariate density estimation requires multiple predictor variables in the model formula, similar to the regression case in Section 3.5. In this section, some examples are presented.

Example 5.5. (Multivariate Density Estimation). The `trimod` dataset is a bivariate dataset with 225 observations from a trimodal distribution. Each of the three components is a bivariate standard normal distribution, with centers at $(3\sqrt{3}/2, 0), (-3\sqrt{3}/2, 3)$ and $(-3\sqrt{3}/2, -3)$. The true peak height is about $1/(6\pi) = 0.053$.

The multivariate density is estimated by specifying multiple terms on the right-hand side of the formula. Here, we fit a local log-quadratic model, with a 35% nearest neighbor bandwidth:

```
> fit.trim <- locfit(~x0+x1, data=trimod, alpha=0.35)
> plot(fit.trim, type="persp")
```

Figure 5.4 shows the fit.

A common density estimation problem is to estimate the smallest region containing a fixed probability mass. At first, constructing such a region may appear to require tricky numerical integration of the density estimate. However, a trick to estimate the contour level is to order the fitted values at the data points, and use the corresponding empirical level.

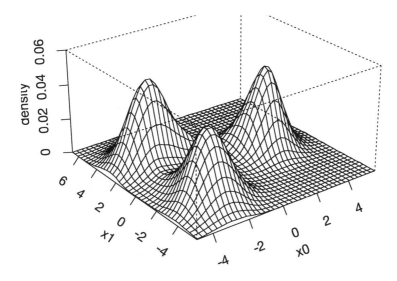

FIGURE 5.4. Bivariate density estimation.

Example 5.6. (Probability Contours). We compute 95% and 50% mass contours for the trimodal sample used in Example 5.5. First, use `fitted` to compute fitted values at the data points. Then, produce a contour plot with the appropriate empirical contour levels:

```
> emp <- sort(fitted(fit.trim))
> plot(fit.trim, vband=F, v=emp[floor(c(0.05,0.5)*225)])
> points(trimod$x0, trimod$x1, col=2, cex=0.5)
```

Figure 5.5 shows the result. The 50% contour defines three separate regions, and the 95% contour has a small hole in the middle.

5.3   Diagnostics for Density Estimation

Does the density estimate fit the data? The question of diagnostics is just as important for density estimation as it is for regression. But answering the question is much more difficult. The source of the problem is simple: There is no natural definition for residuals for a density estimate, and no saturated model. In Section 5.3.1 some possible definitions of residuals are considered, along with graphical displays for detecting lack of fit. Formal goodness of fit criteria based on the likelihood are considered in Section 5.3.2 and squared error methods in Section 5.3.3.

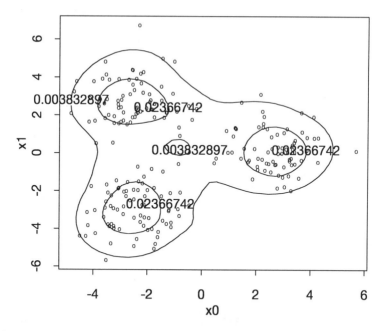

FIGURE 5.5. Probability contour plots: 50% and 95% mass contours for a trimodal example.

5.3.1 Residuals for Density Estimation

There are a number of ways to construct residual type diagnostics for density estimation. Perhaps the most obvious is to compare the integral of the density estimate,

$$\hat{F}(x) = \int_{-\infty}^{x} \hat{f}(u)du,$$

with the empirical distribution function

$$\hat{F}_{emp}(x) = \frac{1}{n}\sum_{i=1}^{n} I(X_i \leq x).$$

Example 5.7. Figure 5.6 shows the empirical distribution function and the integral of a local density estimate. The smoothing parameter for the density estimate is $\alpha = (0.1, 1.2)$, which is larger than that used in Figure 5.1:

```
> fit <- locfit(~geyser, alpha=c(0.1,1.2),
+    flim=c(1,6), renorm=T)
> x <- seq(1, 6, by=0.01)
> z <- predict(fit, x)
```

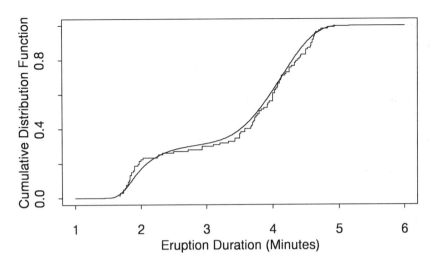

FIGURE 5.6. Empirical distribution function (step curve) and integrated density estimate (smooth curve) for the Old Faithful dataset.

```
> plot(x, 0.01*cumsum(z), type="l")
> lines(sort(geyser), (1:107)/107, type="s")
```

The `renorm=T` argument rescales the density estimate so that it integrates to 1.

In Figure 5.6, the empirical distribution function is steeper than the estimate between 1.8 and 2, which indicates that the peak has been trimmed. The flatness of the empirical distribution function between 2 and 3.5 indicates that the estimate has overfilled the valley.

The P-P and Q-Q plots are based $\hat{F}$ and $\hat{F}_{\text{emp}}$. The P-P (or probability) plot uses the result that $F(X_i)$ behave like a sample from a uniform distribution. If $X_{(i)}$ is the ith order statistic, then $\mathrm{E}(F(X_{(i)})) = i/(n+1)$. Thus, a plot of $\hat{F}(X_{(i)})$ against $i/(n+1)$ should be close to a straight line; large departures from a straight line indicate lack of fit. The Q-Q (quantile) plot transforms back to the observation scale, ploting $X_{(i)}$ against $\hat{F}^{-1}(i/(n+1))$.

An alternative residual diagnostic for density estimation is to begin with a small bandwidth and look at the change in the estimate as the amount of smoothing is increased; can this change be attributed to noise, or does it indicate lack of fit? The simplest implementation of this idea is to begin with a histogram, computed at a small bandwidth. Then, treat the histogram counts and smooth them using local Poisson regression, as described in Section 5.1.3 and Example 5.4. One can then compute residuals for the Poisson model, as discussed in Section 4.3.2.

Example 5.8. We construct residual plots for the Old Faithful geyser dataset. First, a raw histogram of the data is constructed using a bin width of 0.05:

```
> geyser.round <- data.frame(duration=seq(1.05,5.95,by=0.05),
+    count=as.numeric(table(cut(geyser,
+      breaks=seq(1.025,5.975,length=100))))))
```

Note that care is required to ensure zeros are retained. The fit and residual plots can now be constructed:

```
> fit <- locfit(count~duration, data=geyser.round,
+    weights=rep(107*0.05,99), alpha=c(0.1,1.2),
+    family="poisson")
> plot(fit, get.data = T)
> res <- residuals(fit)
> fitr <- locfit.raw(geyser.round$duration, res, alpha=0.1)
> plot(geyser.round$duration, res)
> lines(fitr)
```

Figure 5.7 shows the fits and smoothed residual plots for three different smoothing parameters. As the smoothing parameter decreases, the fit shows the left peak getting sharper and the trough for $2 \leq$ duration ≤ 3.5 getting deeper. The residual plots also show this: In the top residual plot, there is a pronounced peak and five successive positive residuals, around duration $= 1.8$. The residuals also show some evidence of the trough being filled in, even at smallest smoothing parameter.

5.3.2 Influence, Cross Validation and AIC

The likelihood cross validation criterion for density estimation is

$$\mathrm{LCV}(\hat{f}) = \sum_{i=1}^{n} \log \hat{f}_{-i}(X_i) - n \left(\int_{\mathcal{X}} \hat{f}(u)du - 1 \right) \qquad (5.12)$$

where $\hat{f}_{-i}(X_i)$ denotes the density estimate at X_i when this observation is deleted from the dataset. This criterion was first proposed for the kernel density estimate (5.6) by Habbema, Hermans and Van Der Broek (1974) and Duin (1976).

As in Section 4.3.3, the likelihood cross validation score can be approximated using the method of infinitesimal perturbations. This leads to

$$\log \hat{f}_{-i}(X_i) \approx \log \hat{f}(X_i) - \frac{W(0)}{n} e_1^T \mathbf{M}_1^{-1} e_1 + \frac{1}{n} \qquad (5.13)$$

where

$$\mathbf{M}_j = \int_{\mathcal{X}} W \left(\frac{u-x}{h} \right)^j A(u-x)A(u-x)^T e^{\langle \hat{a}, A(u-x) \rangle} du.$$

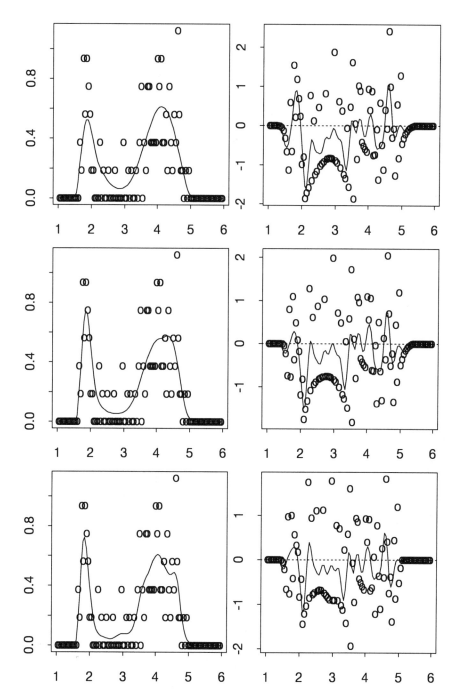

FIGURE 5.7. Fits and smoothed residual plots for geyser data: $\alpha = (0.1, 1.2)$ (top), $\alpha = (0.1, 0.8)$ (middle) and $\alpha = (0.1, 0.5)$ (bottom).

The influence function for density estimation is defined as

$$\text{infl}(x) = n^{-1}W(0)e_1^T\mathbf{M}_1^{-1}e_1; \tag{5.14}$$

the dependence on x is through the matrix $\mathbf{M}_1$. Then

$$\sum_{i=1}^{n} \log \hat{f}_{-i}(X_i) \approx \sum_{i=1}^{n} \log \hat{f}(X_i) - \sum_{i=1}^{n} \text{infl}(X_i) + 1.$$

Summing over the observations leads to the Akaike information criterion for density estimation:

$$\text{AIC}(\hat{f}) = -2\sum_{i=1}^{n} \log \hat{f}(X_i) + 2\sum_{i=1}^{n} \text{infl}(X_i) + 2n\left(\int_{\mathcal{X}} \hat{f}(u)du - 1\right). \tag{5.15}$$

The factor of -2 is introduced here to be consistent with our definition of AIC for local likelihood regression. The quantity

$$\nu_1 = \sum_{i=1}^{n} \text{infl}(X_i)$$

is one definition of the degrees of freedom for a density estimation fit, extending the regression ν_1 defined by (2.16). Correspondingly, we can extend the ν_2 definition to

$$\nu_2 = \sum_{i=1}^{n} \text{vari}(X_i)$$

where $\text{vari}(x) = n^{-1}e_1^T\mathbf{M}_1^{-1}\mathbf{M}_2\mathbf{M}_1^{-1}e_1$.

5.3.3 Squared Error Methods

An entirely different method of cross validation, known as least squares cross validation, was developed for density estimation by Rudemo (1982) and Bowman (1984). This method does not target the likelihood function, but rather the integrated squared error;

$$
\begin{aligned}
\text{ISE}(\hat{f}, f) &= \int_{-\infty}^{\infty} (\hat{f}(x) - f(x))^2 dx \\
&= \int_{-\infty}^{\infty} \hat{f}(x)^2 dx - 2\int_{-\infty}^{\infty} \hat{f}(x)f(x)dx + \int_{-\infty}^{\infty} f(x)^2 dx \tag{5.16}
\end{aligned}
$$

The third term on the right-hand side of (5.16) does not depend on the estimate $\hat{f}(x)$. If the object is to choose $\hat{f}$ to minimize the integrated squared error, then the final term can be ignored. The first term, $\int_{-\infty}^{\infty} \hat{f}(x)^2 dx$, depends only on the density estimate and can be evaluated numerically. The central term can be expressed as

$$\int_{-\infty}^{\infty} \hat{f}(x)f(x)dx = \text{E}(\hat{f}(X))$$

where X is a random variable with density $f(\cdot)$ and is independent of the original sample. This can be estimated by leave-one-out cross validation;

$$\hat{\mathrm{E}}(\hat{f}(X)) = \frac{1}{n} \sum_{i=1}^{n} \hat{f}_{-i}(X_i).$$

This leads to the following definition.

Definition 5.2 The **least squares cross validation** criterion for a density estimate $\hat{f}(x)$ is

$$\mathrm{LSCV}(\hat{f}) = \int_{-\infty}^{\infty} \hat{f}(x)^2 dx - \frac{2}{n} \sum_{i=1}^{n} \hat{f}_{-i}(X_i). \qquad (5.17)$$

As usual, the cross validation component can be approximated using the influence function. Using (5.13) and (5.14), we have

$$\hat{f}_{-i}(X_i) \approx \hat{f}(X_i) \exp(n^{-1}) \exp(-\mathrm{infl}(X_i)) \approx \frac{n}{n-1} \hat{f}(X_i)(1 - \mathrm{infl}(X_i)).$$

Thus, the LSCV criterion can be approximated by

$$\mathrm{LSCV}(\hat{f}) \approx \int_{-\infty}^{\infty} \hat{f}(x)^2 dx - \frac{2}{n-1} \sum_{i=1}^{n} \hat{f}(X_i)(1 - \mathrm{infl}(X_i)).$$

This is exact for local constant fitting.

5.3.4 Implementation

The `aicplot()` and `lcvplot()` functions introduced in Section 4.3.3 can be used directly for density estimation. By default, these ignore the integral term in (5.15). To renormalize the density estimate so that $\int \hat{f}(x)dx = 1$, add the `renorm=T` argument.

The likelihood criteria must be applied rather carefully, since they pay considerable attention to the tail of densities. But any density estimate will perform poorly in the tails and choice of bandwidth is largely an assumption. For example, should a single outlier represent its own little peak in a density, or should it represent a long tail?

Schuster and Gregory (1981) note that LCV, when used to select a constant bandwidth estimate, always selects a bandwidth *larger* than the smallest separation between data points, and thus produces extremely poor results for long tailed distributions. AIC also exhibits anomolous behavior at small bandwidths.

This is not a criticism of AIC or LCV, but simply a recognition that constant bandwidth estimates are poor in tails. The solution comes in two parts. First, ensure that larger bandwidths are used in the tails; for example, by using a nonzero nearest neighbor component in LOCFIT's two-component specification. Second, compare the criteria with the fitted degrees of freedom, and look over a sensible range.

A second problem is caused by ties in the data. This effect has been mostly studied with the LSCV criterion and local constant estimation (Silverman 1986; Sheather 1992). The main result is that if there are too many ties in the data, $\text{LSCV}(\hat{f}_h) \to -\infty$ as $h \to 0$. But again LSCV should not be criticized for this behavior. A sample from a continuous density does not have ties. By selecting $h = 0$, LSCV is simply trying to reproduce the raw data histogram. But problems where this occurs should be treated as discrete, and the LSCV criterion modified accordingly (Exercise 5.4).

Example 5.9. In Figure 5.8 we compute the AIC criterion for local constant, local linear and local quadratic density estimates for the Old Faithful dataset. A typical call to `aicplot()` is:

```
> a0 <- cbind(0.05, c(0.17,seq(0.2,0.7,by=0.05)))
> plot(aicplot(~geyser, alpha=a0, deg=0, renorm=T,
+    flim=c(1,6), ev="grid", mg=51), pch="0")
```

To control tail behavior, the nearest neighbor component of the smoothing parameter is fixed at $\alpha = 0.05$ for local constant and local log-linear fitting, and $\alpha = 0.1$ for local log-quadratic. The constant component h of the smoothing parameter is changed from fit to fit. Corresponding computation of the LSCV criterion is shown on the right of Figure 5.8.

We use the fitted degrees of freedom ν_2 as the x-axis. Both criteria, and each local polynomial degree (0, 1 and 2), show similar patterns. Fewer than five degrees of freedom is inadequate, while for more than five degrees of freedom the criteria are indecisive. Local log-quadratic fitting is better than local log-linear and local constant.

For local quadratic fitting, six degrees of freedom corresponds to the smoothing parameter $(0.1, 0.9)$, and twelve degrees of freedom corresponds to $(0.1, 0.4)$. The AIC criterion relates to what was shown in the fits and residual plots in Figure 5.7. The largest smoothing parameter, $(0.1, 1.2)$ was too large, with little to choose between the smaller parameters.

While all the curves in Figure 5.8 show a similar pattern, the location of the minimum varies substantially. This emphasizes the importance of looking at the whole cross validtaion curve, rather than just the minimum.

If the bandwidths are decreased further, most of the criteria will downturn again, as discreteness and tails of the data take over. But by ploting the criteria against degrees of freedom, as in Figure 5.8, we obtain a sensible view of the data. Fits above 14 degrees of freedom are rarely useful for datasets of 107 observations.

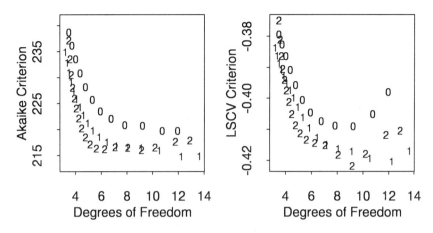

FIGURE 5.8. Akaike's criterion (left) and least squares cross validation (right) for the Old Faithful dataset. Values for local constant fitting (0), local linear fitting (1) and local quadratic fitting (2).

5.4 Some Theory for Density Estimation

This section derives basic theoretical properties for the local likelihood density estimate and develops an approximate distribution theory. The results are similar to the corresponding results for local likelihood regression models in Section 4.4, so only the main ideas are sketched here.

5.4.1 Motivation for the Likelihood

The attractiveness of maximum likelihood estimation stems from (4.12). In the density estimation notation this can be written as

$$E_f \mathcal{L}(f_1) \leq E_f \mathcal{L}(f), \qquad (5.18)$$

with equality only when $f_1 = f$ almost everywhere. With the extended definition of the likelihood (5.1), this property holds for *all* non-negative functions f_1; we do not require f_1 to be a density. One consequence of this extension is that maximum likelihood estimation can be performed with multiplicative parameters. For example, fitting the family $f(x; C, \mu) = C \exp(-(x - \mu)^2/2)$ by maximum likelihood gives $\hat{C} = (2\pi)^{-1/2}$.

The property (5.18) extends to the local log-likelihood;

$$E_f \mathcal{L}(f_1, x) \leq E_f \mathcal{L}(f, x)$$

with equality when $f(u) = f_1(u)$ on the support of $W((u - x)/h)$. This suggests estimating $f(x)$ by maximizing (5.2) over a suitable class of functions.

5.4.2 Existence and Uniqueness

Let C (dependent on the fitting point x, the weight function W and the degree of local polynomial p) be the parameter space:

$$C = \{a = (a_0, \ldots, a_p) : \int_{\mathcal{X}} W\left(\frac{u-x}{h}\right) \exp(\langle a, A(u-x)\rangle)du < \infty\}. \tag{5.19}$$

In many cases the set C is open; for example, if the weight function is bounded and has compact support, $C = \mathcal{R}^d$. In this case, the parameter vector $\hat{a}$ (if it exists) must lie in the interior of C, and it is a solution of the local likelihood equations (5.5).

The Jacobian of the local likelihood (5.3) is

$$J(a) = -\int_{\mathcal{X}} A(u-x)A(u-x)^T W\left(\frac{u-x}{h}\right) \exp(\langle a, A(u-x)\rangle)du.$$

For non-negative weight functions W, this is strictly negative definite. This implies that the local likelihood is concave, and the local likelihood estimate, if it exists, is unique. The following theorem gives precise conditions for existence.

Theorem 5.1 Suppose the parameter space (5.19) is open. The local likelihood density estimate exists if and only if there exists no parameter vector $a_0 \neq 0$ such that

$$\langle a_0, A(X_i - x)\rangle = 0 \quad \forall \quad i : w_i(x) > 0$$

$$\langle a_0, A(u - x)\rangle \leq 0 \quad \forall \quad u : W\left(\frac{u-x}{h}\right) > 0.$$

Proof: Suppose such an a_0 exists. Then

$$\mathcal{L}_x(\lambda e_1 + ca_0) = \lambda \sum_{i=1}^{n} w_i(x) - n \int W\left(\frac{u-x}{h}\right) e^{\lambda + c\langle a_0, A(u-x)\rangle} du.$$

Clearly

$$\lim_{c \to \infty} \mathcal{L}_x(\lambda e_1 + ca_0) = \lambda \sum_{i=1}^{n} w_i(x)$$

$$\lim_{\lambda \to \infty} \lim_{c \to \infty} \mathcal{L}_x(\lambda e_1 + ca_0) = \infty;$$

the likelihood is unbounded and the estimate does not exist.

Conversely, suppose no such a_0 exists. Write

$$\sup_a \mathcal{L}(a, x) = \sup_{a: \|a\|=1} \sup_{\lambda} \mathcal{L}(\lambda a, x); \tag{5.20}$$

we need to show both these suprema are actually achieved. For fixed a with $\|a\| = 1$, we claim (Exercise 5.3)

$$\mathcal{L}_x(\lambda a) = \lambda \sum_{i=1}^{n} w_i(x) \langle a, A(X_i - x) \rangle$$

$$- n \int_{\mathcal{X}} W\left(\frac{u-x}{h}\right) A(u-x) e^{\lambda \langle a, A(u-x) \rangle} du \quad (5.21)$$

is a concave function of λ and tends to $-\infty$ as $\lambda \to \pm\infty$ (or when λ tends to the boundaries of the parameter space $\mathcal{C}$, when this is bounded). Thus the inner supremum of (5.20) must be achieved; let the maximizer be $\lambda = \lambda(a)$. Concavity of $\mathcal{L}(a, x)$ implies $\lambda(a)$ must be continuous on the surface of the unit sphere, and hence the outer supremum is achieved by compactness. □

What does Theorem 5.1 mean in practical terms? For existence of the density estimate, we must be unable to find a polynomial (other than the trivial solution, a constant) that attains its maximum at every point X_i being used in the fit. This generalizes the separating hyperplane theorem for local logistic regression (Exercise 4.4). The local linear estimate exists provided at least one observation has nonzero weight, since a linear function is monotone. A quadratic polynomial may have a single maximum, so the local quadratic estimate exists provided two distinct observations receive nonzero weight.

5.4.3 Asymptotic Representation

The main result of this section is an approximate decomposition of the local likelihood estimate as the sum of a deterministic bias' component and a random component. The result is obtained by linearizing the local likelihood equations, similarly to the techniques used for local likelihood regression in Section 4.4. The following notation is needed:

- $g(x) = \log(f(x))$, and $\tilde{g}$ is the vector of Taylor series coefficients, up to order p.

- $\mathbf{M}_j = \int W(\frac{u-x}{h})^j A(u-x) A(u-x)^T f(u) du; j = 1, 2.$

- $b_p = h^{-(p+1)} \int (u-x)^{p+1} W(\frac{u-x}{h})^j A(u-x) f(u) du.$

- S_n is the left-hand side of the local likelihood equations;

$$S_n = \sum_{i=1}^{n} w_i(x) A(X_i - x). \quad (5.22)$$

The decomposition of the local likelihood estimate is, as $n \to \infty$, $h = h_n \to 0$ and $nh_n \to \infty$:

$$\mathbf{H}(\hat{a} - \tilde{g}) = \frac{h^{p+1} g^{(p+1)}(x)}{(p+1)!} \mathbf{HM}_1^{-1} b_p$$

$$+\frac{1}{n}\mathbf{H}\mathbf{M}_1^{-1}(S_n - \mathrm{E}(S_n)) + o(h^{p+1} + (nh^d)^{-1/2}) \quad (5.23)$$

The first term represents a systematic bias component, and the second term is a random variance component. The bias component as stated is for one dimension; the d-dimensional result requires all partial derivatives of $g(x)$ of order $p+1$. The covariance matrix of S_n is evaluated in Exercise 5.1. A central limit theorem (Loader 1996b) shows asymptotic normality of S_n, and hence of $\hat{a}$. The normal approximation for $\hat{a}$ has the covariance matrix

$$\frac{1}{n}\mathbf{M}_1^{-1}\mathbf{M}_2\mathbf{M}_1^{-1}.$$

By the delta method, the asymptotic variance of $\hat{f}(x)$ is $f(x)^2$ times the $(1,1)$ element of this matrix.

Example 5.10. For the local log-linear density estimate ($p = 1$), one obtains

$$\mathbf{M}_j \approx f(x)\left(\begin{array}{cc} \int W(v)^j\,dv & 0 \\ 0 & h^2\int v^2 W(v)^j\,dv \end{array}\right).$$

This yields the variance and bias approximations

$$\mathrm{E}(\hat{a}_0) - g(x) \approx \frac{h^2}{2}g''(x)\frac{\int v^2 W(v)dv}{\int W(v)dv}$$

$$\mathrm{var}(\hat{a}_0) \approx \frac{1}{nhf(x)}\frac{\int W(v)^2 dv}{(\int W(v)dv)^2}.$$

These can be transformed using the delta method to obtain approximate biases and variances for $\hat{f}(x)$:

$$\mathrm{E}(\hat{f}(x)) - f(x) \approx \frac{h^2}{2}f(x)g''(x)\frac{\int v^2 W(v)dv}{\int W(v)dv}$$

$$\mathrm{var}(\hat{f}(x)) \approx \frac{f(x)}{nh}\frac{\int W(v)^2 dv}{(\int W(v)dv)^2}.$$

5.5   Exercises

5.1 Consider S_n defined by (5.22), where $X_1, \ldots, X_n$ are independent identically distributed random variables with density $f(x)$.

a) Show

$$\mathrm{E}(S_n) = n\int W\left(\frac{u-x}{h}\right)A(u-x)f(u)du$$

$$\mathrm{cov}(S_n) = n\mathbf{M}_2 - \frac{1}{n}\mathrm{E}(S_n)\mathrm{E}(S_n)^T.$$

Derive a similar expression for the covariance matrix $\text{cov}(S_n)$.

b) Suppose the density is continuous at x with $f(x) > 0$, $n \to \infty$, $h = h_n \to 0$ and $nh \to \infty$. Let $\mathbf{H}$ be as defined in (2.37). Show

$$\frac{1}{\sqrt{n}} \text{cov}(\mathbf{H}^{-1} S_n) = f(x) \int W(v)^2 A(v) A(v)^T dv + o(1);$$

in particular, the covariance term involving $\text{E}(S_n)$ is asymptotically negligible. Evaluate $n^{-1} E(\mathbf{H}^{-1} S_n)$ using a Taylor series for $f(x)$, retaining terms up to $o(h^2)$.

c) Using Chebycheff's inequality show, on a componentwise basis, $P(|\mathbf{H}^{-1}(S_n - E(S_n))| \geq \epsilon) \to 0$ for all $\epsilon > 0$. Hence, show

$$\mathbf{H}^{-1} S_n \to f(x) \int W(v) A(v) dv \tag{5.24}$$

in probability.

d) Using (5.24) and the local likelihood equations, show

$$\mathbf{H}\hat{a} \to \begin{pmatrix} \log f(x) \\ 0 \\ \vdots \\ 0 \end{pmatrix}$$

in probability and that the local likelihood density estimate is consistent.

5.2 Consider local log-quadratic density estimation in d dimensions, using the Gaussian weight function.

a) Write down the local likelihood equations. Express the right-hand side in terms of the multivariate integrals

$$\int W\left(\frac{u}{h}\right) e^{a + b^T u + u^T \mathbf{C} u} du;$$

$$\int u W\left(\frac{u}{h}\right) e^{a + b^T u + u^T \mathbf{C} u} du; \tag{5.25}$$

$$\int u u^T W\left(\frac{u}{h}\right) e^{a + b^T u + u^T \mathbf{C} u} du. \tag{5.26}$$

Here, b is a vector in $\mathcal{R}^d$ and $\mathbf{C}$ is a symmetric $d \times d$ matrix.

b) Show

$$\int W\left(\frac{u}{h}\right) e^{a + b^T u + u^T \mathbf{C} u} du$$

$$= (2\pi)^{d/2} \exp(a + \frac{1}{2} b^T \mathbf{M}^{-1} b) \det(\mathbf{M})^{-1/2}$$

where $\mathbf{M} = h^{-2} I - 2\mathbf{C}$. Derive closed forms for (5.25) and (5.26).

c) Provide a closed form solution for the density estimate. What condition is necessary for existence of the estimate? Is the parameter space open?

5.3 Consider the log-likelihood $\mathcal{L}_x(\lambda a, x)$ with fixed a, $\|a\| = 1$. Suppose a does *not* satisfy the conditions of the vector a_0 in Theorem 5.1. That is, either $\langle a, A(X_i - x)\rangle \neq 0$ for some i with $w_i(x) > 0$ or $\langle a, A(u - x)\rangle$ has both positive and negative regions on the support of $W((u - x)/h)$. Show that $\mathcal{L}_x(\lambda a) \rightarrow -\infty$ as $\lambda \rightarrow \pm\infty$.

5.4 Izenman and Sommer (1988) and Sheather (1992) have fitted kernel density estimates to the postage stamp data (Example 5.4) using the Gaussian kernel and standard deviation about 0.0013. In LOCFIT terms, this is a constant bandwidth of $2.5 \times 0.0013 = 0.00325$.

a) Evaluate and plot this fit. Compare with the local log-quadratic fit (Figure 5.3) and the data. Is the kernel estimate adequate for modeling the peaks?

b) Develop an LSCV algorithm for discrete Poisson regression for kernel density estimation. Use the loss function $\sum_{i=1}^{n}(\hat{p}_i - p_i)^2$ where p_i is the probability of the ith bin. The cross validation should use leave-one-observation-out; not leave-one-bin-out. Consider the behavior of LSCV(h) at small bandwidths. In particular, show it has a finite limit as $h \rightarrow 0$ (Bonus: Use the influence function; don't restrict to `deg=0`).

c) Write an S function to evaluate the discrete LSCV criterion using a LOCFIT fit. Apply this function to the postage stamp data. Compare with the results of Sheather (1992).

Remark. The point of this exercise is that discrete data does not have densities, and this is particularly important for model selection when small bandwidths are used.

6
Flexible Local Regression

In this chapter we look at the flexibility that can be obtained by changing the components of local regression: the coefficients, the fitting criteria and the weight functions. The specific problems studied include:

- Higher order coefficients and local slopes (section 6.1).

- Periodic and seasonal smoothing (Section 6.2).

- One-sided smoothing and discontinuous function estimation (Section 6.3).

- Robust local regression (Section 6.4).

6.1 Derivative Estimation

Derivatives are of natural interest in many settings. At the most basic level, the derivative $\mu'(x)$ measures the effect of the independent variable x on the mean response. In particular, $\mu'(x) = 0$ implies the covariate is having no effect.

As emphasized in Section 6.1.1, the problem of derivative estimation is plagued by identifiability and interpretation difficulties. To make any real headway, one must be willing to *assume* that if the local polynomial fits the data within the smoothing window, then the local slope provides a good approximation to the derivative. This leads to the following local slope estimate.

Definition 6.1 Let $\hat{a} = (\hat{a}_0 \quad \cdots \quad \hat{a}_p)$ be the local coefficient estimates from a local polynomial fit (2.5), with $p \geq 1$. The **local slope estimate** is

$$\hat{\mu}'(x) = \langle \hat{a}, A'(0) \rangle = \hat{a}_1. \tag{6.1}$$

The local slope estimate (6.1) is not the derivative of the fitted curve $\hat{\mu}(x)$. The exact derivative is (Exercise 6.1)

$$\frac{d\hat{\mu}(x)}{dx} = \hat{a}_1 + e_1^T (\mathbf{X}^T \mathbf{W} \mathbf{X})^{-1} \mathbf{X}^T \mathbf{W}' \hat{\epsilon} \tag{6.2}$$

where $\hat{\epsilon}$ is the vector of residuals from the local polynomial,

$$\hat{\epsilon} = (\mathbf{I} - \mathbf{X}(\mathbf{X}^T \mathbf{W} \mathbf{X})^{-1} \mathbf{X}^T \mathbf{W}) Y,$$

and $\mathbf{W}'$ is the derivative of $\mathbf{W}$, with diagonal elements $dw_i(x)/dx$. For a constant bandwidth $h(x) = h$,

$$\frac{dw_i(x)}{dx} = -\frac{1}{h} W' \left(\frac{x_i - x}{h} \right).$$

The exact derivative (6.2) is clearly more work to compute than the local slope (6.1). Usually, little is gained by the exact computation, since the derivative can be represented as a local slope estimate by a modification of Henderson's theorem.

Theorem 6.1 The weight diagram for the local slope estimate (6.1) has the form

$$l_i(x) = w_i(x) \langle \alpha, A(x_i - x) \rangle$$

and satisfies the reproducing equations

$$\sum_{i=0}^{n} P(x_i - x) l_i(x) = P'(0) \tag{6.3}$$

where $P(u)$ is any polynomial of degree $\leq p$. Conversely, if a weight diagram satisfies (6.3) and has at most p sign changes, then it is a local slope estimate of degree p.

6.1.1 Identifiability and Derivative Estimation

Before discussing the implementation of derivative estimation, we need to consider carefully what is required. The mathematical definition of the derivative $\mu'(x)$ of a function $\mu(x)$ is

$$\mu'(x) = \lim_{\delta \to 0} \frac{\mu(x + \delta) - \mu(x)}{\delta}, \tag{6.4}$$

provided the limit exists.

Now consider the local regression setting with equally spaced observations $x_i = i/n$, and the mean function is $E(Y_i) = \mu(x_i)$. Consider the family of functions

$$\mu_a(x) = \mu(x) + \frac{a}{2\pi n} \sin(2\pi nx).$$

At the data points, these functions have the following property:

$$
\begin{aligned}
\mu_a(x_i) &= \mu(x_i) + \frac{a}{2\pi n} \sin(2\pi i) \\
&= \mu(x_i).
\end{aligned}
$$

That is, the mean function, evaluated at the data points, is independent of a. The likelihood function does not depend on a, and thus the dataset provides absolutely no way of distinguishing among the class of functions $\mu_a(x)$.

Now consider differentiation:

$$
\begin{aligned}
\mu_a'(x_i) &= \mu'(x_i) + a\cos(2\pi i) \\
&= \mu'(x_i) + a.
\end{aligned}
$$

While the functions $\mu_a(x)$ are indistinguishable, the derivatives are quite different. By changing a, the distribution of the observations does not change in any way, but the derivative could be anything! Thus there can be no reasonable data-based estimate of the derivative, and any estimate has to make strong assumptions about the true mean function.

Why are derivatives of interest? Do derivatives ever exist in practice? For example, the velocity of a moving object can be defined as the derivative of position with respect to time. But it is far from clear whether the limit (6.4) exists: Is position of an object even defined well enough (at microscopic levels) to make this determination?

What's really of interest is a local linear approximation: Find a constant c such that

$$\mu(x + \delta) \approx \mu(x) + \delta c$$

for a sufficient range of values of δ to be useful. Differentiation is a convenient mathematical formulation of this: If $\mu(x)$ is differentiable, $c = \mu'(x)$ may provide a satisfactory value of c. But what is really required in practice is a local slope: Fit a local linear (or higher order) model, and consider the local slope coefficient.

At first reading, the distinction between derivatives and local slopes may seem rather pedantic. But the distinction has important consequences, particularly in problems involving bias estimation. A plug-in estimate of the bias of a local regression estimate substitutes an estimate for the derivatives appearing in (2.34) or (2.41). But for noisy data, the derivative estimate is highly dependent on the bandwidth: As the amount of smoothing increases, the derivative estimates decrease in magnitude. In practice, any such estimate amounts to an assumption that a local polynomial approximation to $\mu(x)$ also provides a good approximation to the derivatives. See Exercise 6.3. These problems have particular relevance to confidence interval construction (Section 9.2) and bandwidth selection (Chapter 10).

Does asymptotic theory, showing consistency and other properties of derivative estimates, resolve these issues? Simply put, no, since the results provide no information about the accuracy of the estimates. More precisely, one can derive bias approximations, similar to (2.34), for derivative estimates. But these in turn depend on higher order derivatives, and trying to estimate these leads to circular arguments.

Does this discussion apply to estimation of the mean $\mu(x)$ itself? Largely, it does; formally, the mean function $\mu(x)$ cannot be estimated anywhere *except* at the data points, unless formal smoothness assumptions are made. Looking closely at the goodness of fit statistics developed earlier, such as cross validation and CP, we see these formally only measure goodness of fit at the data points. A good fit at the data points tells us nothing about how well the fit performs between data points; rather, we *assume* the true mean is smooth, and thus fitting between data points is reasonable.

6.1.2 *Local Slope Estimation in* LOCFIT

By default, LOCFIT returns the local fit. To obtain the local slope, one gives the `deriv` argument to the `locfit()` call. The value of `deriv` can be either a variable name or number. Thus,

```
> fit <- locfit(NOx~E+C, data=ethanol, deriv=2)
```

and

```
> fit <- locfit(NOx~E+C, data=ethanol, deriv="C")
```

both produce the fit for the derivative with respect to C. To obtain higher order derivatives, use `deriv=c("C","C")` for the second derivative with respect to `"C"`, or `deriv=c("C","E")` for the mixed derivative. The order of the derivative cannot exceed the order of the local polynomial fit.

Derivatives for local likelihood estimates are found in the same manner. However, one must remember that the local polynomial is fitted through the link function $\theta(x) = g(\mu(x))$, and no attempt is made to back-transform the coefficients.

Example 6.1. We estimate the derivative of the density $f(x)$ of the Old Faithful dataset. By default, the estimation is via the log link, $g(x) = \log(f(x))$. Since $f'(x) = g'(x)f(x)$, we can estimate $f'(x)$ by estimating both the density $f(x)$ and the derivative $g'(x)$:

```
> fit1 <- locfit(~geyser, alpha=c(0.1,0.6), flim=c(1,6))
```

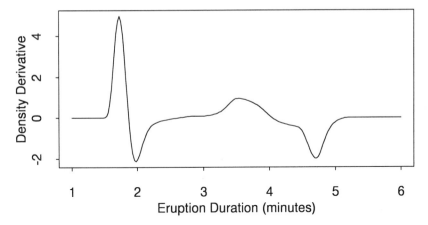

FIGURE 6.1. Estimating the density derivative for the Old Faithful data.

```
> fit2 <- locfit(~geyser, alpha=c(0.1,0.6), flim=c(1,6),
+    deriv=1)
> z <- lfmarg(fit1, 200)
> plot(preplot(fit1,z) * preplot(fit2,z))
```

Note the computation of `fit1` (for the density) and `fit2` (for the derivative of the log-density). A grid (`z`) of 200 points is generated, and the predictions are computed on this grid. The resulting plot is shown in Figure 6.1.

6.2   Angular and Periodic Data

Angular data arises when measurements are made on a circle. Periodic data frequently arises when data are measured over time: for example, there may be a daily or annual pattern in the data.

Figure 6.2 shows a series of monthly measurements of carbon dioxide (CO_2) measurements at the Mauna Loa observatory from 1959 to 1990 (Boden, Sepanski and Stoss (1992)). There are two obvious components in the data. First, there is an overall increasing trend, approximated by the solid line in Figure 6.2. Second, there is the cyclical component, with a peak in May every year and trough in October.

Suppose one is interested in predicting future carbon dioxide measurements. Simply smoothing the data is not satisfactory. A fit with a large bandwidth as shown in Figure 6.2 misses the periodic structure. A local quadratic smooth with a bandwidth small enough to capture the annual peaks and troughs would be better, but would not capture the regularity of

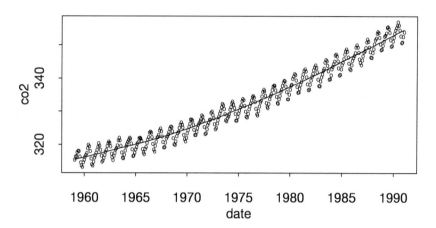

FIGURE 6.2. Carbon dioxide measurements at Mauna Loa observatory. Super-imposed is a long-term trend, estimated by a local linear smooth.

the periodic component and would provide poor predictions of future CO_2 levels.

To adequately model this dataset, periodicity must be enforced in the smooth. This can be achieved by enforcing periodicity in the weight function. A periodic, or circular, distance function is defined as

$$d(x_1, x_2) = 2|\sin((x_1 - x_2)/(2s))| \qquad (6.5)$$

where s is a scale parameter. Some properties of this distance function include:

- If $x_1 = x_2$, then $d(x_1, x_2) = 0$.

- If $x_1 - x_2$ is small, then $\sin((x_1 - x_2)/(2s)) \approx (x_1 - x_2)/(2s)$, and so $d(x_1, x_2) \approx |x_1 - x_2|/s$.

- If $x_1 - x_2 = \pi s$, then $d(x_1, x_2) = 2$, the maximum value.

- If $x_1 - x_2 = 2\pi s$, then $d(x_1, x_2) = 0$. The distance function is periodic with period $2\pi s$.

The smoothing weights are defined in the usual manner:

$$w_i(x) = W\left(\frac{d(x_i, x)}{h}\right).$$

A circular local quadratic model

$$\mu_x(x_i) = a_0 + a_1 s \sin((x_i - x)/s) + a_2 s^2 (1 - \cos((x_i - x)/s))$$

is fitted, using the weights $w_i(x)$, to get local parameter estimates $\hat{a}_0, \hat{a}_1, \hat{a}_2$. The smooth estimate is $\hat{\mu}(x) = \hat{a}_0$. The basis functions are defined so that $\hat{a}_1$ estimates the local slope and $\hat{a}_2$ estimates the local curvature.

Example 6.2. Suppose the carbon dioxide dataset can be expressed as the sum of trend and periodic components:

$$Y_i = \mu(x_i) + \nu(m_i) + \epsilon_i \tag{6.6}$$

where $\mu(x_i)$ is the long-term trend, m_i is the month of the ith observation and $\nu(m_i)$ is the periodic component. This is an additive model and can be fitted using the backfitting algorithm. But since the data extends over many periods, the x_i and m_i components are nearly orthogonal, and a single iteration will suffice. First, the long-term trend is estimated using a local linear smooth and a large span:

```
> fit1 <- locfit(co2~I(year+month/12), data=co2, alpha=0.5,
+   deg=1)
```

The resulting fit was shown in Figure 6.2. A periodic smooth is then fitted to the residuals of this fit:

```
> loc.co2 <- cbind(co2,res = residuals(fit1))
> fit2 <- locfit(res~ang(year+month/12), data=loc.co2,
+   scale=1/(2*pi), alpha=c(0,2))
> plot(fit2, xlim=c(0,1))
```

The angular term `ang(year+month/12)` in the model formula specifies a periodic smooth. The periodic component has a period of one year, so the scale parameter s in (6.5) must be set to $1/2\pi$ years. This is achieved through the `scale=1/(2*pi)` argument.

The estimated periodic component is shown in Figure 6.3. This shows the asymmetry of the periodic component: The decline from peak to trough takes about five months, whereas the rise from trough to peak takes about seven months. Thus, just using a sine wave for the periodic component would be unsatisfactory. The smoothing parameter can be a little hard to visualize with angular data; the effective bandwidth is approximately $hs = 2/(2\pi) = 0.31$ in the preceeding example. Also useful is the fitted degrees of freedom; the angular component for this fit has $\nu_2 = 4.15$.

Figure 6.4 shows the sum of the long-term trend and periodic components. This is computed simply by adding fitted values for both fits:

```
> plot(co2$year+co2$month/12, fitted(fit1)+fitted(fit2),
+   type="l")
```

The additive model (6.6) assumes the periodic component remains constant over time. But this may not be reasonable; for example, we might expect the amplitude of the seasonal fluctuation to increase as the overall

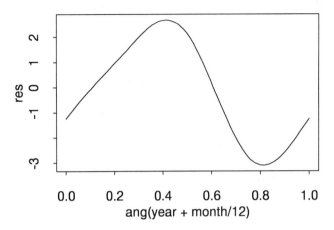

FIGURE 6.3. Carbon dioxide data: Smooth estimate of the cyclical component.

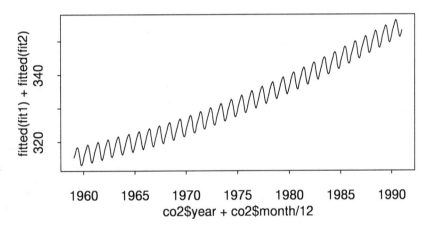

FIGURE 6.4. Fitted values, adding the long term trend and cyclical components.

trend increases. An alternative local analysis for periodic components, suggested by Cleveland (1993), is to split the dataset into the twelve monthly series, and smooth each month separately. This has the advantage of allowing some trends in the seasonal component. The disadvantage is that smoothness of the seasonal trend is not allowed for.

An alternative is to fit the periodic component through a bivariate model, where one variable is used to represent a long-term trend, and a second term is used to represent the periodic component. This has the advantage of allowing the periodic component to change slowly over time.

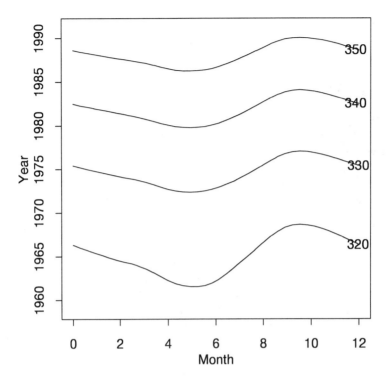

FIGURE 6.5. Carbon dioxide data set. Fitting the trend and periodic component as a bivariate model.

Example 6.3. A bivariate model is fitted to the CO_2 dataset. The first predictor is month, fitted with a periodic component. The second predictor is year+month/12, fitted with straight local regression:

```
> fit <- locfit(co2~ang(month)+I(year+month/12), data=co2,
+    scale=c(12/(2*pi),10),alpha=0.2)
> plot(fit)
```

Figure 6.5 shows the fit. Of course, all the data points lie on a series of diagonal lines, beginning at the bottom left and ending at the top right.

A major interpretation difficulty in Example 6.3 is the scale variable; what are the relative amounts of smoothing in the two variables? The distance in this case is

$$d((y_1, m_1), (y_2, m_2))^2 = (2\sin((m_1 - m_2) \cdot \pi/12))^2 + \left(\frac{y_1 - y_2}{10}\right)^2.$$

Thus, 10 years in the y direction is about the same as $12/(2\pi) = 1.9$ months in the m direction.

6.3 One-Sided Smoothing

In some situations, we may like to smooth using only the data before, or only the data after, a fitting point x. One example is the detection of discontinuities in the fitted curve. Another example is in the forecasting of time series and in particular cross validation for this problem (Li and Heckman 1996 and Exercise 6.6).

The problem of modeling discontinuous curves has attracted considerable attention in recent years. Two classes of algorithms can be found in the literature. First, smoothing algorithms can attempt to preserve discontinuities in curves, without formally identifying the points of discontinuity. Running medians are a simple example of this type of algorithm. A more sophisticated method is that of McDonald and Owen (1986), which uses cross validation methods to choose between one- and two- sided smoothers. Locally adaptive smoothing algorithms, such as wavelets (Donoho and Johnstone 1994), and local bandwidth rules discussed in Chapter 11 and references therein, fall into this category.

The second class of algorithms are two-stage procedures. First, one estimates the points of discontinuity. Second, the data is smoothed between the points of discontinuity, either by explicitly splitting the dataset, or by choice of basis functions to preserve the discontinuities. Examples of algorithms of this type include Lee (1989), Müller (1992), Loader (1996a), Jose and Ismail (1997) and Qiu and Yandell (1998).

The first stage, identification of the change points, is performed using one-sided smoothers. Specifically, for a fitting point t, left and right weight functions are defined as

$$w_i^-(t) = \begin{cases} W\left(\frac{t-x_i}{h}\right) & x_i < t \\ 0 & \text{otherwise} \end{cases}$$

$$w_i^+(t) = \begin{cases} W\left(\frac{x_i-t}{h}\right) & x_i \geq t \\ 0 & \text{otherwise} \end{cases}.$$

In particular, note the left weights $w_i^-(t)$ give nonzero weights only to data points $x_i < t$, and the right weights $w_i^+(t)$ only to points $x_i \geq t$. These smoothing weights are used to fit local polynomial models, leading to one-sided estimates $\hat{\mu}_-(t)$ and $\hat{\mu}_+(t)$ respectively.

The estimates $\hat{\mu}_-(t)$ and $\hat{\mu}_+(t)$ can be interpreted as estimates of the left and right limits of $\mu(x)$, as $x \to t$. If $t = \tau$ is a point of discontinuity, these limits are different, and

$$\hat{\Delta}(\tau) = \hat{\mu}_+(\tau) - \hat{\mu}_-(\tau)$$

is an estimate of the jump size. We expect the process $\hat{\Delta}(t)$ to be peaked when t is near a discontinuity τ, and close to 0 when t is not near a discontinuity. The estimate of τ is simply the maximizer:

$$\hat{\tau} = \text{argmax}_t \hat{\Delta}(t)^2.$$

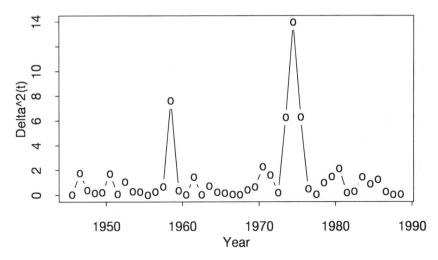

FIGURE 6.6. Change point estimation: Estimating $\hat{\Delta}(t)^2$.

Multiple, and well separated, peaks of the process $\hat{\Delta}(t)^2$ can be considered an indication of multiple discontinuities. Since the two estimates are formed with non-overlapping sequences of observations, $\text{var}(\hat{\Delta}(t)) = \text{var}(\hat{\mu}_-(t)) + \text{var}(\hat{\mu}_+(t))$, which gives some indication as to whether peaks represent real discontinuities. But this ignores the multiple comparison nature of the problem. Also, peaks in $\hat{\Delta}(t)$ can be caused by sharp curvature in the mean $\hat{\mu}(t)$ rather than discontinuities. There is no completely satisfactory solution of these problems in the literature.

Example 6.4. Scott (1992, page 234) discusses a dataset measuring the thickness of U.S. pennies. The dataset consists of the thickness of two coins for each year from 1945 to 1989. During this time period, there were (at least) two changes in the production thickness, see Figure 6.7. To detect the changes, one-sided left and right smooths are fitted to the data:

```
> midp <- (1945:1988)+0.5
> fitl <- locfit(thickness~left(year), data=penny,
+   alpha=c(0,10), deg=1, ev=midp)
> fitr <- locfit(thickness~right(year), data=penny,
+   alpha=c(0,10), deg=1, ev=midp)
> plot((preplot(fitr)-preplot(fitl))^2, type="b")
```

The one-sided smooths are obtained using `left(x)` and `right(x)` in the model formula. Note that one-sided smooths are discontinuous, so LOCFIT's default method of interpolating fits from a sparse set of points is undesirable in this case. Thus the fits are computed at the midpoints (`midp`) between successive years.

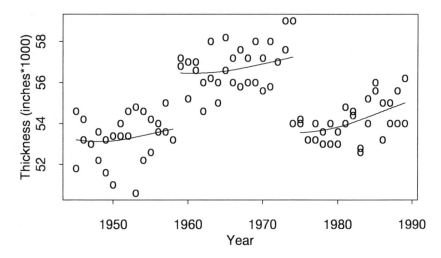

FIGURE 6.7. Split smoothing of the penny dataset.

The estimate $\hat{\Delta}(t)^2$ is shown in Figure 6.6. Two changes, $t = 1958.5$ and $t = 1974.5$, are clearly indicated by the sharp peaks.

After points of discontinuity have been estimated, one obtains the final fitted curve by splitting the data into pieces, and applying local regression to each segment. This can be achieved simply using either the xlim or subset arguments.

Example 6.5. We continue with the penny data, using the change point estimates $\hat{\tau}_1 = 1958.5$ and $\hat{\tau}_2 = 1974.5$. A local linear fit is computed for the data for $t \le 1958$:

```
> fit0 <- locfit(thickness~year, data=penny,
+    alpha=c(0,10), deg=1, subset=(year<=1958))
```

Fits fit1 and fit2 are computed similarly, using data $1959 \le t \le 1974$ and $t \ge 1975$, respectively. The fits are superimposed on a plot of the data:

```
> plot(penny$year, penny$thickness)
> lines(fit0)
> lines(fit1)
> lines(fit2)
```

Figure 6.7 shows the three segments. There is also some evidence of a gradual increase of the mean thickness within segments.

Remark. Careful inspection of Figure 6.7 suggests that the 1974 coins should be split, one before the change and one after the change.

6.4   Robust Smoothing

When the errors ϵ_i have a long tailed distribution, the local least squares estimate can be overly sensitive to extreme observations. Robust regression methods attempt to remedy this by identifying and downweighting influential observations. Chapter 6 of Hampel, Ronchetti, Rousseeuw and Stahel (1986) contains an extensive review of robust methods for parametric regression. The most widely used methods are based on M-estimation.

There are several ways to adapt M-estimation to local regression. One scheme was proposed as part of the LOWESS procedure of Cleveland (1979). The algorithm used by Cleveland is:

1. Assign all observations a robustness weight $v_i = 1$.

2. Smooth the data using local polynomials, with weights $v_i w_i(x)$. That is, the product of the robustness weights and localization weights.

3. Compute the residuals $\hat{\epsilon}_i = Y_i - \hat{\mu}(x_i)$, and estimate the scale s as the median of the absolute values of the residuals.

4. Assign observations robustness weights

$$v_i = B(\hat{\epsilon}_i/s)$$

where $B(\,\cdot\,)$ is a robustness weight function; Cleveland uses $B(z) = (1 - z^2/6)_+^2$.

5. Repeat steps 2, 3 and 4 until convergence.

The robustness arises from the downweighting at the fourth step. An observation with $\hat{\epsilon}_i = 0$ receives robustness weight 1, while an observation with $|\hat{\epsilon}_i| > \sqrt{6}s$ receives robustness weight 0.

A difficulty with the straightforward implementation occurs at small bandwidths. Suppose the smoothing window $(x - h(x), x + h(x))$ for some fitting point x contains less than half the data. Since the scale estimate is defined as the median absolute residual, up to half the data can be rejected as outliers. In extreme cases, this could mean all the observations in $(x - h(x), x + h(x))$ are rejected. This point was noted by Maechler (1992).

Katkovnik (1979, 1985) and Tsybakov (1986) proposed robust versions of local regression by changing the fitting criterion. For a symmetric nonnegative function $\rho(\,\cdot\,)$ with $\rho(0) = 0$, the local M-estimate is obtained by maximizing the criterion

$$\mathcal{L}_x(a) = -\sum_{i=1}^{n} w_i(x)\rho\left(\frac{Y_i - \langle a, A(x_i - x)\rangle}{s}\right). \qquad (6.7)$$

This is similar to the local log-likelihood (4.2), and algorithms and theory for M-estimation closely follow those for local likelihood. In particular, if

	$\rho_1(v)$	$\rho_2(v)$	$\rho_3(v)$	$\rho_4(v)$
Normal	1.000	1.571	1.313	1.052
Double Exp.	2.000	1.000	1.179	1.434
Cauchy	∞	2.467	2.000	3.521

TABLE 6.1. Robustness variance factors for three densities and four robustness functions: $\rho_1(v) = v^2/2$, $\rho_2(v) = |v|$, $\rho_3(v) = -\log(1 + x^2)$ and $\rho_4(v)$ defined by (6.8) with $c = 2$.

$\rho(v)$ is convex, the existence and uniqueness results of Theorem 4.1 continue to hold. A common choice is

$$
\begin{aligned}
\rho_c(v) &= x^2 I(|v| < c) + (2c|v| - c^2)I(|v| \geq c) \qquad (6.8) \\
\rho_c'(v) &= -2cI(v \leq -c) + 2vI(|v| < c) + 2cI(v \geq c) \\
\rho_c''(v) &= 2I(|v| < c)
\end{aligned}
$$

where c is a prespecified constant. Huber (1964) showed this had a certain minimax optimality property.

For a fixed dataset, the resulting estimate converges to the local least squares estimate as $c \to \infty$ and to the local L_1 estimate (Katkovnik 1985, section 7.3; Wang and Scott 1994) as $c \to 0$.

6.4.1 Choice of Robustness Criterion

A theoretical development of robust local M-estimation is found in Tsybakov (1986), following earlier developments of Stuetzle and Mittal (1979) and Tsybakov (1982) in the local constant case. The ideas largely parallel those for parametric regression models, developed by Huber (1964, 1981) and Hampel, Ronchetti, Rousseeuw and Stahel (1986), among others.

If the errors ϵ_i have a symmetric density $g(v)$, the variance of the estimate $\hat{\mu}(x)$ is approximately

$$
\text{var}(\hat{\mu}(x)) \approx \frac{\int \dot{\rho}(v/s)^2 g(v) dv}{(\int \ddot{\rho}(v/s)g(v)dv)^2} \|l(x)\|^2. \qquad (6.9)
$$

This follows from a derivation similar to the local likelihood variance, (4.19). The optimal choice of ρ is $\rho(v) = -\log(g(v))$; in this case, the local robust regression becomes local likelihood estimation.

Since the density $g(v)$ is treated as unknown, one must choose a function $\rho(v)$ that exhibits good behavior over some class of densities $g(v)$. For $\rho(v) = v^2$, (6.9) reduces to $\sigma^2 \|l(x)\|^2$, which is poor for heavy tails. Choosing $\rho(v) = |v|$ leads to $\|l(x)\|^2/(4g(0)^2)$, which is poor when $g(0)$ is small, such as for a bimodal density. The choice (6.8), with a robust estimate for the scale parameter, is a compromise between these two extremes.

Table 6.1 summarizes the variance factors for the normal, double exponential and Cauchy distributions, and four choices of $\rho(v)$.

6.4.2   Choice of Scale Estimate

The choice of scale estimate has a critical effect on the performance of robust estimates. The most commonly used scale estimates are based on the absolute values of the residuals ϵ_i; for example,

$$\hat{s} = \text{median}|\hat{\epsilon}_i|.$$

Here, the median is taken over all observations with nonzero weights.

Any scale s will produce a set of local parameters $\hat{a} = \hat{a}(s)$ and correspondingly a set of residuals $\hat{\epsilon}_i = Y_i - \langle \hat{a}(a), A(x_i - x) \rangle$ and a new scale estimate $\hat{s}(s)$. The robust M-estimate iterates this process, until a stable scale and set of parameters are found. Thus, we solve the equation

$$\hat{s}(s) - s = 0.$$

This equation always has a solution, since $\hat{s}(0) \geq 0$ and $\hat{s}(\infty) < \infty$. But the solution need not be unique (even for convex $\rho(\cdot)$), and occasionally highly nonrobust solutions can be found.

An alternative scale estimate[1] is to treat the robustness criterion as a log-likelihood, with s as a scale parameter. Thus, (6.7) is replaced by

$$\sum_{i=1}^{n} w_i(x) \left(\rho \left(\frac{Y_i - \langle a, A(x_i - x) \rangle}{s} \right) - \log(s) \right),$$

which can be simultaneously maximized over a and s.

For variance estimation and confidence interval construction, another scale estimate is required, specifically, of the lead factor in (6.9). The estimate used is

$$\hat{\sigma}^2 = \frac{n}{n - 2\nu_1 + \nu_2} \frac{\sum_{i=1}^{n} \dot{\rho}(\hat{\epsilon}_i/s)^2}{\sum_{i=1}^{n} \ddot{\rho}(\hat{\epsilon}_i/s)}.$$

6.4.3   LOCFIT Implementation

LOCFIT supports both the global reweighting and local M estimation algorithms. The `locfit.robust()` function implements the global reweighting algorithm; see Section B.2.2 for more details. This has the advantage that the robustness reweighting is performed entirely in S-Plus, so the downweighting function $B(z)$ can easily be chosen by the user.

Local M estimation is supported through the families `family="huber"` and `family="cauchy"` without scale estimation, and the corresponding quasi families `"qhuber"` and `"qcauchy"` for M estimation with scale estimation.

---

[1]Suggested to the author by Xuming He.

6.5   Exercises

6.1 Working from (2.23), derive the exact derivative (6.2).

Hint. To avoid differentiating $\mathbf{X}$, initially center the fitting functions around a fixed point x_0, instead of the fitting point x. Since the answer must be independent of x_0, you can set $x_0 = x$ *after* doing the differentiation.

6.2 Consider local quadratic density with the identity link (5.7) and $w_j(x) = \phi(x_j - x)$ where $\phi(\cdot)$ is the standard normal density. Show

$$\begin{pmatrix} \hat{a}_0 \\ \hat{a}_1 \\ \hat{a}_2 \end{pmatrix} = \frac{1}{n} \sum_{i=1}^{n} \begin{pmatrix} \frac{1}{2}(3 - (x_i - x)^2) \\ x_i - x \\ (x_i - x)^2 - 1 \end{pmatrix} \phi(x_i - x).$$

Evaluate the derivatives $d\hat{a}_0/dx$ and $d^2\hat{a}_0/dx^2$; in particular, show these do not equal $\hat{a}_1$ and $\hat{a}_2$. Obtain expressions for the coefficients $(\hat{b}_0, \hat{b}_1, \hat{b}_2, \hat{b}_3)$ for a local cubic fit, and compare with the derivatives of the local quadratic fit.

6.3 For the Old Faithful geyser dataset (Example 5.2), estimate the local slope using local log-quadratic density estimation. Compute an estimate of $\int_1^6 f'(x)^2 dx$. Plot the estimate as a function of the smoothing parameter. Perform the exercise for regression examples (such as the ethanol dataset) and for second derivatives.

6.4 The object of this exercise is to split the CO_2 dataset into two components (1959-1986 and 1987-1990) and use models fitted to the first component to predict the CO_2 readings in the second component.

a) Fit the first component using a local linear smooth with nearest neighbor span of 0.4. Use predict() to predict the fitted values in the second component. Compute the mean squared prediction error.

Note: To split the dataset into two separate components, use

```
> co2a <- co2[co2$year<=1986,]
> co2b <- co2[co2$year>=1987,]
```

In your locfit() call, set flim=c(1959,1991) to ensure that the fit points span the prediction points.

b) Compute the residuals from the fit, and fit a periodic smooth to the residuals. Compute the predictions based on adding the long-term and periodic smooths. Compute the mean squared prediction error. Make a plot of the 1987 to 1990 data, and superimpose the two sets of predicted values.

c) Compute separate smooths for each month with the 1959 to 1986 data, and use these smooths to predict the corresponding months in the 1987 to 1990 data. Again, compute the mean squared prediction error.

Note: To subset the data for smoothing, use, for example,

```
> locfit(co2~year,data=co2a,subset=(month==1),...)
```

where `subset` specifies which months to use, and `...` represents whatever other arguments you decide to use.

d) Compute a bivariate smooth, similar to Figure 6.5 for the `co2a` dataset. Again, compute the mean squared prediction error for the `co2b` points. Can the results be improved, either by changing the `scale` in the `year` direction, or by changing `alpha`?

6.5 Suppose data (x_i, Y_i) is equally spaced, with $x_i = i/n$ and $Y_i = \mu(x_i) + \epsilon_i$. The errors ϵ_i are $N(0,1)$ and the mean function $\mu(t)$ has a discontinuity at an unknown $\tau \in (0,1)$. We estimate τ using locally constant fitting with one-sided rectangular kernels. Let $k = nh$ where h is the bandwidth.

a) Ignoring edge effects at 0 and 1, show

$$k(\hat{\Delta}(\tau) - \hat{\Delta}(\tau + j/n)) = \sum_{i=1}^{j} (2Y_{n\tau+i} - Y_{n\tau+k+i} - Y_{n\tau-k+i}).$$

b) Suppose as $n \to \infty$, $k/n \to 0$ and $k/\log(n) \to \infty$. Show

$$n(\hat{\tau} - \tau) \Rightarrow \operatorname{argmax}(\sqrt{1.5}Z_i - \frac{1}{2}|i\Delta|) \qquad (6.10)$$

where $\{Z_i\}_{i=-\infty}^{\infty}$ is a two sided random walk of $N(0,1)$ random variables (The condition $k/\log(n) \to \infty$ is necessary, for reasons discussed in Section 13.1).

Remark. If $W(0) > 0$ and W is continuous on $[0, \infty)$, Loader (1996a) showed that (6.10) holds *without* the factor $\sqrt{1.5}$. This matches results of Hinkley (1970) for change point estimation in parametric models.

6.6 Given a time series $Y_1, \ldots, Y_n$, suppose we wish to use local polynomial methods to forecast one step ahead. The object is to choose the bandwidth and degree to minimize the forecast mean squared error.

a) Explain why cross validation, as developed in Chapter 2, is inappropriate for this problem.

b) Develop a one-sided cross validation algorithm, forecasting each observation Y_i from $Y_{i-h}, \ldots, Y_{i-1}$.

c) Implement this algorithm using LOCFIT's one-sided smoothing. Apply the algorithm to Spencer's mortality dataset.

7
Survival and Failure Time Analysis

The problem of survival analysis arises in medical statistics: Observations T_i may represent the time a patient dies following a treatment or the length of time the patient remains infected with a disease. A closely related problem arising frequently in industrial statistics is that of failure times: The observations T_i represent the times to failure of a component.

A distinguishing feature of many survival and failure time datasets is censoring. Patients in a study may be lost to followup, so the survival time is not known. In failure time studies, components may still be working at the conclusion of the study.

In this chapter a number of ways of analyzing censored survival data using local likelihood models are presented. The most basic problem is to characterize the distribution of survival times. This is closely related to the density estimation problem in Chapter 5. But in survival analysis, it is often more convenient to focus on the hazard rate, which represents the conditional probability of death at time t, given that a subject has survived up to time t. The problem is defined more formally, and estimation methods discussed, in Section 7.1.

Often, one wants to relate the distribution of survival times to a covariate. For example, do patients receiving a treatment tend to survive longer? This hazard regression problem can be treated in several ways. Section 7.1.4 discusses estimation of the conditional hazard rate. This procedure requires large datasets to work well, and informative results can often be obtained using simpler models. In Section 7.2 regression methods are used to estimate the conditional mean survival time. Section 7.3 extends these methods to likelihood models.

7.1   Hazard Rate Estimation

The problem of estimating a smooth hazard rate has a long history, dating to Watson and Leadbetter (1964) who developed some versions of kernel estimates. In this section, we develop a local likelihood approach to hazard rate estimation, based on the methods of Wu and Tuma (1990) and Hjort (1993).

7.1.1   Censored Survival Data

Suppose a study involves n patients, with survival times $T_1, \ldots, T_n$. Each patient is associated with a censoring time c_i and the observations are

$$Y_i = \begin{cases} T_i & T_i < c_i \\ c_i & T_i \geq c_i \end{cases}.$$

It is always known whether an observation is censored ($Y_i = c_i$) or uncensored ($Y_i < c_i$).

Suppose the survival times T_i have density $f(t)$ and distribution function $F(t) = P(T_i \leq t)$. The observations Y_i have density $f(y)$ for $y < c_i$, and a discrete mass $P(Y_i = c_i) = 1 - F(c_i)$. Frequently, interest is not in $f(t)$ directly, but rather in the hazard rate, which represents the probability of death at time t, given that the patient has survived until time t. Formally, the hazard rate $h(t)$ of a random variable T is defined as

$$h(t) = \lim_{\delta \to 0} \delta^{-1} P(T \leq t + \delta | T \geq t).$$

One can easily show that

$$h(t) = \frac{f(t)}{1 - F(t)}. \tag{7.1}$$

The density can also be expressed in terms of the hazard rate; integrating (7.1) gives

$$-\int_{-\infty}^{t} h(u)du = [\log(1 - F(u))]_{-\infty}^{t} = \log(1 - F(t))$$

and hence

$$1 - F(t) \;=\; \exp\left(-\int_{-\infty}^{t} h(u)du\right)$$

$$f(t) \;=\; h(t) \exp\left(-\int_{-\infty}^{t} h(u)du\right). \tag{7.2}$$

More discussion of hazard rates and detailed derivations of the properties presented here can be found in books such as Miller (1981) and Cox and Oakes (1984).

7.1.2   The Local Likelihood Model

The likelihood for censored observations can be written in terms of the hazard function from (7.2):

$$\prod_{\substack{i=1 \\ Y_i < c_i}}^{n} f(Y_i) \prod_{\substack{i=1 \\ Y_i = c_i}}^{n} (1 - F(Y_i)) = \prod_{\substack{i=1 \\ Y_i < c_i}}^{n} h(Y_i) \exp\left(-\sum_{i=1}^{n} \int_{-\infty}^{Y_i} h(u) du\right). \quad (7.3)$$

Note the product is taken over only the uncensored observations, while the sum is taken over all observations, both censored and uncensored. We remark that work in survival analysis often assumes that the c_i are random variables and includes appropriate terms in the likelihood. Since our interest is in the survival times, we treat censoring times as nuisance parameters and condition on their values.

The likelihood is locally weighted, and a local polynomial approximation for $\log(h(u))$ leads to the local log-likelihood at a time t:

$$\begin{aligned}
\mathcal{L}_t(a) &= \sum_{i=1}^{n} W\left(\frac{Y_i - t}{h}\right) \langle a, A(Y_i - t) \rangle \\
&\quad - \sum_{i=1}^{n} \int_{-\infty}^{Y_i} W\left(\frac{u - t}{h}\right) e^{\langle a, A(u-t) \rangle} du \\
&= \sum_{i=1}^{n} W\left(\frac{Y_i - t}{h}\right) \langle a, A(Y_i - t) \rangle \\
&\quad - \int_{-\infty}^{\max(Y_i)} N(u) W\left(\frac{u - t}{h}\right) e^{\langle a, A(u-t) \rangle} du \quad (7.4)
\end{aligned}$$

where $N(u)$ is the number of observations at risk at time t; $N(u) = \sum_{i=1}^{n} I(Y_i \geq u)$. The local likelihood estimate of the hazard rate is

$$\hat{\lambda}(t) = e^{\langle \hat{a}, A(0) \rangle}$$

where $\hat{a}$ is the vector maximizing (7.4). Note that the local likelihood for hazard rates is similar to the likelihood (5.3) for density estimation, but with the factor $N(u)$ in the integral.

Example 7.1. The local constant $(p = 0)$ hazard rate estimate is

$$\hat{\lambda}(t) = \frac{\frac{1}{nh} \sum_{\substack{i=1 \\ Y_i < c_i}}^{n} W\left(\frac{Y_i - t}{h}\right)}{\frac{1}{nh} \sum_{i=1}^{n} \int_{-\infty}^{Y_i} W\left(\frac{u - t}{h}\right) du}. \quad (7.5)$$

If $W(\cdot)$ is the rectangular weight function, then (7.5) is the number of deaths in $[t - h, t + h]$, divided by the at-risk time in $[t - h, t + h]$ summed over all observations.

There is an alternative interpretation. Without censoring, the numerator of (7.5) is simply the kernel density estimate (5.6), and the denominator is the kernel-smoothed empirical cumulative distribution function. Thus,

$$\hat{\lambda}(t) = \frac{\hat{f}(t)}{\frac{1}{h} \int_{-\infty}^{\infty} W\left(\frac{u-t}{h}\right)(1 - \hat{F}(u))du}, \qquad (7.6)$$

giving a natural estimate of (7.1).

Now suppose censoring times are drawn independently from a distribution with c.d.f. $G(c)$. In this case, the numerator of (7.5) estimates $f(t)(1 - G(t))$ (termed the subdensity by Antoniadis, Grégoire and Nason (1999)), and the denominator estimates $(1 - F(t))(1 - G(t))$. Again, we have a natural estimate of the hazard rate.

7.1.3 Hazard Rate Estimation in LOCFIT

The local likelihoods for density estimation and hazard rate estimation are very similar, and implementation of the hazard rate estimate just requires the use of a special integration routine. But there are a number of other issues that need to be considered.

The most serious problem occurs at the right boundary, especially when the largest observation is uncensored. At this point, we have just one observation, and it dies immediately. Thus, an empirical estimate would say that death is certain at this time, corresponding to an infinite hazard rate. This effect is also observed in smooth hazard rate estimates; for example, in (7.6), $\hat{F}(u) = 1$ at the largest observation, so the denominator is near 0. This effect can lead to distracting tails on the plots of hazard rate estimates.

To reduce the visual distraction, the hazard rate estimate implemented in LOCFIT uses a modified form of the local likelihood (7.4), replacing $N(u)$ by $N(u) + 1$. This effectively adds a censored observation to the dataset, equal to the largest real observation. This is mainly a visual fix; since there is usually little data to estimate the hazard rate in the right tail, the estimate will still be highly variable, and one should beware of reading too much into the estimate.

Hazard rate estimation in LOCFIT is specified with `family="hazard"`. An indicator variable showing whether or not the ith observation is censored can be provided as the `cens` argument. Other options for density estimation are also applicable to hazard rate estimation. In particular, the identity link (`link="ident"`) uses local polynomial approximations for $\lambda(t)$ rather than $\log(\lambda(t))$. As with density estimation, this is largely equivalent to higher order kernel estimates, such as those proposed by Müller and Wang (1994).

Example 7.2. The Stanford heart transplant dataset, from Miller and Halpern (1982), reports survival times of 184 heart transplant recipients. Of these patients, 113 died during the followup period and 71 were either

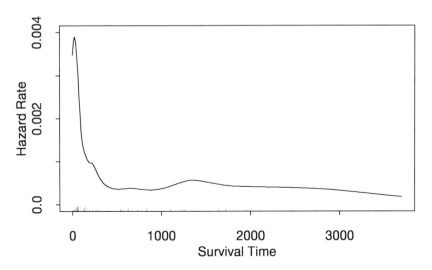

FIGURE 7.1. Local likelihood hazard estimation applied to heart transplant data

still alive when the data was collected or otherwise lost to followup. The hazard rate is estimated using a local log-quadratic model, with $\alpha = 0.4$:

```
> fit <- locfit(~surv, data=heart, cens=cens, alpha=0.4,
+    family="hazard", xlim=c(0,100000))
> plot(fit, mpv=300, ylim=c(0,0.004), xlab="Survival Time",
+    ylab="Hazard Rate", get.data=T)
```

The censoring variable can be either logical or 0-1, with 1 or TRUE indicating a censored observation.

The plot in Figure 7.1 shows an initially high hazard rate (up to 0.004, or one death in 250 at-risk days), followed by a sharp drop and leveling off at one death in 2000 days. A previous analysis by Loader (1991), using change point models, also showed a precipitous drop around 60 days.

An important point to note is the provision of the xlim argument. In common with density estimation, the observations are assumed (by default) to have unbounded support, and the integral of (7.4) is evaluated with lower limit $-\infty$. In practice, survival times will nearly always have a support bounded below by 0, and the xlim argument incorporates this constraint. The large upper limit is effectively ∞.

7.1.4 Covariates

Much of the survival analysis literature considers models with covariates, addressing (in many different ways) the question of how the covariates affect survival times. One approach is to estimate the conditional hazard rate $h(t|x)$ for a set of covariates x. Kooperberg, Stone and Truong (1995) fit this model using regression spline models for the log hazard rate, while Gray (1996) obtains estimates by discretizing the data and using local regression. The problem can be addressed directly using the local likelihood approach.

Assuming the hazard function varies smoothly as a function of both time t and covariates x, a local likelihood model is fitted by localizing on both t and x. The local likelihood (7.4) is modified to

$$
\mathcal{L}_{t,x}(a) = \sum_{i=1}^{n} W\left(\frac{Y_i - t}{h}, \frac{x_i - x}{h}\right) \langle a, A(Y_i - t, x_i - x)\rangle
$$
$$
- \sum_{i=1}^{n} \int_{-\infty}^{Y_i} W\left(\frac{u - t}{h}, \frac{x_i - x}{h}\right) e^{\langle a, A(u-t, x_i-x)\rangle} du
$$

where $A(t, x)$ is the vector of fitting functions; for local quadratic fitting with a single covariate, $A(t, x) = (1 \quad t \quad x \quad t^2 \quad xt \quad x^2)^T$. $W(t, x)$ is the weight function; for example, the spherically symmetric weights are $W(t, x) = W_0(\sqrt{t^2 + \|x\|^2})$ for a one dimensional weight function $W_0(\cdot)$.

Example 7.3. We use the Liver metastases dataset from Haupt and Mansmann (1995) and also studied in Antoniadis, Grégoire and Nason (1999). The dataset measures survival times for 622 patients diagnosed with liver metastases, of which 261 are censored. The analysis of hazard rates in Antoniadis, Grégoire and Nason (1999) shows an initially low hazard rate, increasing to about 0.05 at $t = 20$ months, then slowly declining.

We fit the hazard regression model, using diameter of the metastases as the covariate:

```
> fit <- locfit(~t+dm, data=livmet, cens=1-z, scale=0, deg=1,
+    family="hazard", alpha=0.5, xlim=list(t=c(0,10000)))
> plot(fit, ylab="Diameter (c.m.)",
+    xlab="Survival Time (Months)", get.data=T)
```

When `family="hazard"`, the first term in the formula is interpreted as the survival time, and the remaining terms are covariates. The `xlim` argument is used to set a lower bound for the survival times.

The result, in Figure 7.2, shows that the increase in hazard rate is nonuniform. For small diameters, the hazard remains roughly constant. The increasing hazard is most pronounced for larger diameters.

7.2 Censored Regression

Unless large quantities of data are available, full estimation of the hazard rate in the presence of covariates may be an unrealistic goal. Often it is

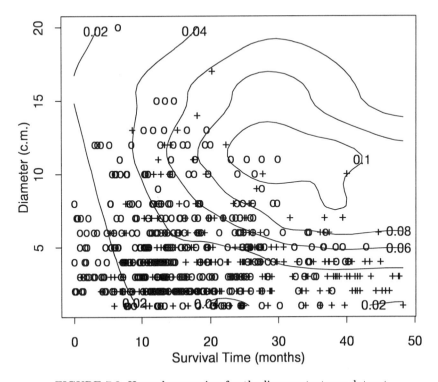

FIGURE 7.2. Hazard regression for the liver metastases dataset.

sufficient to study covariate effects: Does a treatment increase survival time, and by how much? The proportional hazards model is widely used:

$$\lambda(t|x) = e^{\beta(x)}\lambda_0(t). \tag{7.7}$$

Cox (1972) proposed the partial likelihood method for estimating covariate effects, assuming a parametric model for $\beta(x)$. Tibshirani and Hastie (1987) study local versions of partial likelihood.

The partial likelihood algorithm treats the baseline hazard $\lambda_0(t)$ as a nuisance parameter and only provides estimates of the covariate effects. An alternative approach, used by Volf (1989) and Gentleman and Crowley (1991), is to alternate between estimating the covariate effect and the baseline hazard rate.

In this section the proportional hazards model is not assumed, but instead we estimate the conditional mean function $E(T_i|x_i)$ for censored data. The starting point is a regression model with additive errors. The essential idea is to replace each censored observation with a guess of the true observation and then apply local regression to construct a mean estimate.

The guess can be constructed using either parametric or nonparametric methods.

In Section 7.3, the methods are extended to local likelihood estimates. For survival data, this approach is often more realistic than the additive error model.

7.2.1   Transformations and Estimates

The censored regression model assumes observations (x_i, Y_i, δ_i) where Y_i are censored survival times and δ_i are censoring indicators. The additive error model is

$$T_i = \mu(x_i) + \epsilon_i \tag{7.8}$$

where the T_i are the uncensored (and unobserved) survival times. The errors ϵ_i have mean 0, density $f(t)$ and distribution function $F(t)$. The object is to estimate the mean function $\mu(x)$. Since survival times are positive, (7.8) would normally be assumed for a transformation, such as the log of the survival times.

In the absence of censoring, (7.8) is the ordinary local regression model. In the presence of censoring, $Y_i = \min(T_i, c_i)$ where c_i is the censoring time for the ith individual. We would like to replace Y_i by

$$Y_i^* = \mathrm{E}(T_i | Y_i = y, c_i) = \begin{cases} y & y < c_i \\ \mathrm{E}(T_i | T_i \geq y) & y = c_i \end{cases}. \tag{7.9}$$

That is, uncensored observations are left unchanged, while censored observations are replaced by the best guess of their true value. The unconditional density of T_i is $f(t - \mu_i)$, and the *mean residual life* $\mathrm{E}(T_i | T_i \geq y) - y$ is found by integrating the tail of this density:

$$\mathrm{E}(T_i | T_i \geq y) - y = \frac{1}{1 - F(y - \mu_i)} \int_y^\infty (t - y) f(t - \mu_i) dt. \tag{7.10}$$

If $\epsilon_i \sim N(0, \sigma^2)$, it is fairly easy to derive the explicit expression

$$\mathrm{E}(T_i | T_i \geq y) = \mu_i + \sigma \frac{\phi((y - \mu_i)/\sigma)}{1 - \Phi((y - \mu_i)/\sigma)}. \tag{7.11}$$

This model can be fitted by alternating least squares, first proposed by Schmee and Hahn (1979) for parametric regression models. First, smooth the raw data (x_i, Y_i) to obtain an initial estimate of $\mu(x)$ and σ. Then alternately estimate Y_i^* using (7.11) and smooth (x_i, Y_i^*). This procedure is iterated until convergence. A variance estimate is also needed; at each iteration, we compute

$$\hat{\sigma}^2 = \frac{1}{n_u - 2\mathrm{tr}(\mathbf{L}) + \mathrm{tr}(\mathbf{L}^T \mathbf{L})} \sum_{i=1}^n (Y_i - \hat{\mu}(x_i))(Y_i^* - \hat{\mu}(x_i)),$$

where n_u is the number of uncensored observations. Motivation for this estimate is given in Exercise 7.4; note that it reduces to (2.18) in the absence of censoring.

7.2.2 Nonparametric Transformations

An objection to the preceding iterative scheme is that the transformation (7.11) is heavily dependent on the normality assumption. This assumption will be difficult to check, since in the presence of censoring the estimated residuals $Y_i^* - \hat{\mu}(x_i)$ are not normally distributed, even asymptotically.

Can Y_i^* be estimated using less parametric methods? Often, censoring results from termination of a study and the censoring times c_i are bounded. In this case, there is no information about the tails of the distribution of the Y_i, and hence Y_i^* cannot be estimated consistently. Obtaining good estimates without a fully parametric model for the residual distribution - and determining precisely under what conditions such estimates work - is a difficult problem. A rather technical literature has been developed for this problem with a parametric form for $\mu(x)$; see, for example, Lai and Ying (1991) and references therein.

In a smooth regression model, Fan and Gijbels (1994) provide the first use of nonparametric transformations for censored observations, although their proposed schemes have several other deficiencies (Exercise 7.2).

An alternative scheme, proposed by Buckley and James (1979) in the context of a linear model for $\mu(x)$, is to express the mean residual life in terms of the distribution function $F(t - \mu(x_i))$ and substitute a nonparametric estimate $F(\cdot)$. Integration by parts shows the mean residual life (7.10) is

$$\mathrm{E}(T_i | T_i \geq y) - y = \frac{\int_y^\infty (1 - F(t - \mu(x_i)))dt}{1 - F(y - \mu(x_i))}. \qquad (7.12)$$

A suitable nonparametric estimate is the product limit estimate of Kaplan and Meier (1958). This generalizes the empirical distribution function in the presence of censoring. Suppose $U_1, \ldots, U_n$ are censored observations from a distribution $F(u)$, and let $N(u)$ be the number of at-risk observations at time u;

$$N(u) = \sum_{i=1}^n I(U_i \geq u).$$

Assuming there are no ties in the data, the Kaplan-Meier estimate assigns the probability mass $P(U = u | U \geq u) = 1/N(u)$ at each *uncensored* observation $U_i = y$ and 0 everywhere else. This conditional mass function leads to the cumulative distribution function estimate

$$\hat{P}(U_i > u) = 1 - \hat{F}(u) = \prod_{\substack{i=1 \\ U_i \leq u}}^n \left(1 - \frac{I(U_i < c_i)}{N(u)}\right).$$

In the regression setting, the Kaplan-Meier estimate is applied to the estimated residuals $U_i = Y_i - \hat{\mu}(x_i)$. The resulting distribution function estimate is substituted into (7.12) to estimate of Y_i^*. This procedure is iterated, alternately estimating the mean $\hat{\mu}(x)$ and distribution function $F(u)$.

Other nonparametric transformations are possible. If the censoring times c_i are assumed to be random variables with distribution $G(c|x) = P(c_i < c)$, a transformation proposed by Koul, Sursala and Van Ryzin (1981) is

$$Y_i^* = \begin{cases} \frac{Y_i}{1-G(Y_i|x_i)} & Y_i < c_i \\ 0 & Y_i = c_i \end{cases} . \tag{7.13}$$

Leurgans (1987) proposed

$$Y_i^* = \int_0^{Y_i} \frac{1}{1 - G(y|x_i)} dy. \tag{7.14}$$

Fan and Gijbels (1994) considered linear combinations of these two transformations. Under some regularity conditions (in particular, that the censoring distribution is unbounded), these transformations are easily shown to be unbiased; $E(Y_i^*) = T_i$. On the other hand, the transformations have the unintuitive behavior of transforming the uncensored observations and inherently depend on the stochastic assumptions for the censoring times.

Example 7.4. We apply the censored regression model to the Stanford heart transplant dataset, using the age of the patient at transplant as a predictor variable. Since this dataset is highly skewed and bounded below by 0, it makes little sense to fit the model (7.8) directly. We first transform the survival times, considering the response $\log(\Delta + \mathbf{surv})$. The shift $\Delta = 0.5$ avoids $\log(0)$ problems. Three fits are computed: using the raw data and ignoring censoring, correction based on the normal model and correction based on the Kaplan-Meier estimate.

The code follows. The censored regression iterations are performed using the `locfit.censor` function; for a description of how this works, see Section B.2.2. The `km=T` argument uses the Kaplan-Meier estimate of mean residual life; otherwise, the normal model is used.

```
> plotbyfactor(heart$age, 0.5+heart$surv, heart$cens,
+    ylim=c(0.5,16000), log="y")
> fit <- locfit(log10(0.5+surv)~age, data=heart)
> lines(fit, tr = function(x)10^x)
> fit <- locfit.censor(log10(0.5+surv)~age, cens=cens,
+    data=heart, lfproc=locfit.censor)
> lines(fit, lty=2, tr = function(x)10^x)
> fit <- locfit.censor(log10(0.5+surv)~age, cens=cens,
+    data=heart, km=T)
> lines(fit, lty=3, tr=function(x)10^x)
```

Figure 7.3 shows the fits. Both the normal and Kaplan-Meier models make a substantial correction for censoring over the left half of the plot.

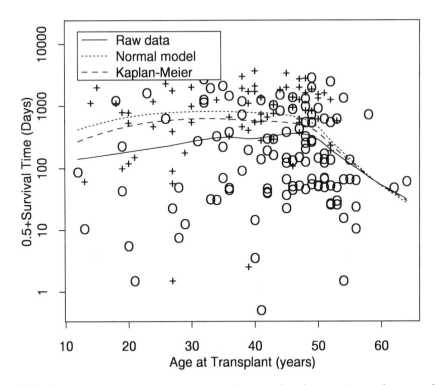

FIGURE 7.3. Stanford heart transplant dataset. Local regression and censored regression using normal and Kaplan-Meier models.

However, the correction made by the normal model is larger than that made by the Kaplan-Meier model. We might expect the Kaplan-Meier correction to be inadequate for this data, since nearly all the largest survival times are censored, and thus the estimates of the mean residual life for these observations are likely to be downward biased. But given the substantial extrapolation there is no real way to validate which - if either - of the estimates is making the right correction.

7.3   Censored Local Likelihood

The regression models in the previous section assume additive errors. But in many datasets, survival times are positive and the additive error model is unreasonable. An alternative is to fit censored local likelihood models. If the responses T_i have density $f(t, \theta_i)$ then (7.3) generalizes to

$$\mathcal{L}(\theta) = \sum_{i=1}^{n} l(Y_i, \theta_i, c_i)$$

where

$$l(Y_i, \theta_i, c_i) = \begin{cases} \log(f(Y_i, \theta_i)) & Y_i < c_i \\ \log(1 - F(Y_i^-, \theta_i)) & Y_i = c_i \end{cases} ;$$

$f(Y_i, \theta_i)$ and $F(Y_i, \theta_i)$ are the density and distribution function of Y_i respectively. This likelihood can be localized and estimates of θ obtained following Definition 4.1.

Most of the methods developed for local likelihood in Chapter 4 can be extended to censored local likelihood. Most importantly, the local likelihood estimate continues to be a solution of the local likelihood equations (4.14). The nice behavior of the solution characterized in Theorem 4.1 continues to hold in many important cases.

For most common families, $\log(1 - F(Y_i^-, \theta_i))$ is a monotone function of θ_i and does not tend to $-\infty$ at one boundary. However, it is bounded above by 0, so the existence part of the theorem will continue to hold if $\mathbf{WX}$ has full rank after censored observations are deleted from $\mathbf{X}$. Concavity (and hence uniqueness) can be checked in special cases. The developments of cross validation, the influence function and AIC in Section 4.3.3 extend to censored models, although the form of $\ddot{l}(Y_i, \theta_i, c_i)$ is quite different for censored observations.

Example 7.5. For the geometric distribution and the log link,

$$P(Y_i \geq y) = q^y = \frac{e^{\theta_i y}}{(1 + e^{\theta_i})^y}.$$

Thus the censored log-likelihood is

$$l(Y_i, \theta, c_i) = \begin{cases} \theta Y_i - (Y_i + 1) \log(1 + e^{\theta_i}) & Y_i < c_i \\ \theta Y_i - Y_i \log(1 + e^{\theta_i}) & Y_i = c_i \end{cases} .$$

For censored observations with $Y_i = c_i$, $l(Y_i, \theta, c_i) \to 0$ as $\theta_i \to \infty$, so that these observations violate the conditions of Theorem 4.1. The second derivative is

$$\frac{\partial^2}{\partial \theta^2} l(Y_i, \theta, c_i) = -\frac{(Y_i + I(Y_i < c_i))e^\theta}{(1 + e^\theta)^2}.$$

This is negative for any $Y_i > 0$, so concavity is preserved. Thus the existence and uniqueness of Theorem 4.1 continue to hold, provided only uncensored observations with $Y_i > 0$ are counted when determining the rank.

Results similar to Example 7.5 hold for other common likelihoods, although establishing concavity tends to be messy. The Poisson family with both the log and identity links is studied in Exercise 7.5.

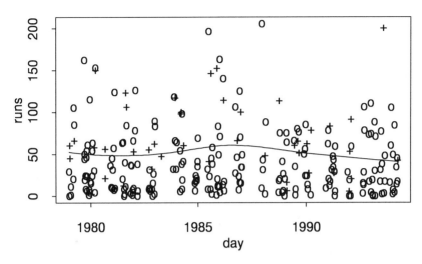

FIGURE 7.4. Batting record of Australian cricketer Allan Border. Circles represent completed (out) innings, and '+' represents censored (not out) innings.

7.3.1 Censored Local Likelihood in LOCFIT

Censored regression is currently supported in LOCFIT for the Gaussian, Poisson, geometric and gamma families by providing a **cens** argument.

Example 7.6. The sport of cricket is played between two teams. Like baseball, players on the batting team attempt to score runs, while players on the fielding team attempt to get batsmen[1] 'out'; for example, by catching a ball hit by a batsman. Unlike baseball, a batsman continues batting, and accumulating runs, until he gets out. The total number of runs scored represents a completed innings for the batsman.

Occasionally, a batsman will not be able to complete his innings. For example, the team may have scored sufficient runs to win the match, so there is no point in continuing. In this case, the innings is recorded as 'not out', which is treated as a censored observation. Figure 7.4 displays the batting record between 1977 and 1994 played by Australian cricketer Allan Border. The dataset contains 265 innings, with a total of 11174 runs. Of these innings, 44 were 'not out' (censored).

There is considerable interest in measuring the performance of players, and the most common measure of a batsman's performance is batting average, defined as the total number of runs scored, divided by the number

---

[1] or batswomen.

of completed innings. For the dataset in Figure 7.4, the average is

$$\frac{11174}{265 - 44} = 50.56.$$

Of course, innings played in 1977 have little relevance to performance in 1990, so we would like to localize this computation. A natural model is the geometric distribution: At any time, we suppose a batsman has a probability $1/(1 + \mu)$ of getting out and a probability $\mu/(1 + \mu)$ of scoring another run. Using the date of the innings as a covariate, we can now fit a local likelihood model:

```
> fit0 <- locfit(runs~day, cens=no, data=border,
+    family="geom", alpha=0.7)
> plot(fit0,get.data=T)
```

The fitted curve is shown in Figure 7.4. The fitted curve here is quite flat, with perhaps a slight peak around 1986. This represents a remarkably consistent performance over the 18 year period.

The geometric model might be criticized in the preceding example, since no allowance is made for several sources of variation, such as strength of the opposing side or condition of the playing field. Thus, the data may be overdispersed, and exhibit more variability than the geometric model would predict. This can be judged by looking at the total deviance (`-2*fit@dp["lk"]`), which in this case is 429.6. Since this is much larger than the sample size (265), the evidence is that the data is overdispersed.

In Section 4.3.4, we considered quasi-likelihood as a method for handling overdispersion. But this strategy is not useful for censored models, since the tail probabilities $F(Y_i^-, \theta_i, c_i)$ are needed, and specifying a relation between the mean and variance functions is not sufficient.

An alternative is to fit a larger family of models. The negative binomial family (4.6), with shape parameter n_i, has a discrete hazard rate

$$P(Y_i = y | Y_i \geq y) \to 1 + \mu(x_i)$$

as $y \to \infty$. When $n_i < 1$, the hazard rate decreases to this asymptote, and the distribution of Y_i is overdispersed, relative to the geometric distribution. This type of behavior has been observed in cricket scores previously by Kimber and Hansford (1993) and is fairly consistent with the dataset used in example 7.6. See exercise 7.6.

Example 7.7. We fit the negative binomial model using a global constant model for the shape parameter and a local quadratic model for the mean function. The shape parameter can be estimated by maximum likelihood. There is no automated way to do this; the easiest way is to compute the local fit for a few candidate values and carefully reconstruct the log likelihood using the **dnbinom** and **pnbinom** functions for the uncensored and censored observations respectively. The following code computes the likelihood for the shape parameter $w = 0.8$:

```
> w <- 0.8
> fit <- locfit(runs~day, cens=no, weights=rep(w,265),
+    data=border, family="geom", alpha=0.7)
> fv <- fitted(fit)
> phat <- fv/(1+fv)
> lk <- rep(0,265)
> no <- border$no==1
> lk[no] <- 1 - pnbinom(border$runs[no]-1,
+    size=w, prob=1-phat[no])
> lk[!no] <- dnbinom(border$runs[!no], size=w,
+    prob=1-phat[!no])
> sum(log(lk))
[1] -1085.228
```

For this example, $\hat{w} = 0.8$ was found to be the maximum likelihood estimate. The final fit is obtained with

```
> fit1 <- locfit(runs~day, data=border, weights=rep(0.8,265),
+    cens=no, family="geom", alpha=0.7)
```

Of particular interest is the difference between the mean estimate under the negative binomial model and the mean estimate under the geometric model. This difference can be plotted by

```
> plot(0.8*preplot(fit1) - preplot(fit0))
```

Figure 7.5 shows the result. Since the geometric model has the 'no-memory' property, this provides an estimate of the effect of censoring on the batsman's average. From this fit, we estimate that the batting average would have been about 1.5 runs higher, had all innings been played to completion.

The magnitude of the difference in Figure 7.5 may be sensitive to the negative binomial assumption. But the sign of the difference is a consequence of a decreasing hazard rate (Exercise 7.3), and does not depend on the particular model used. This is contrary to other models: Table 3 of Kimber and Hansford (1993) suggests that means should be decreased to compensate for censoring.

In the preceding example, the dispersion parameters n_i have been modeled as the global constant w. A more sophisticated approach would be to use local models for both the mean and dispersion parameters. See, for example, Nelder and Pregibon (1987) and Rigby and Stasinopoulos (1996).

Another distribution commonly used for survival times is the Weibull model, with densities

$$f(t, a, b) = \frac{bt^{b-1}}{a} \exp(-t^b/a).$$

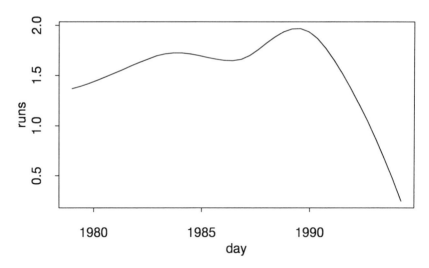

FIGURE 7.5. Batting data: Estimating the censoring effect on the average.

This model is mathematically easier to work with than the negative binomial and gamma distributions. Some properties are:

- The mean is
$$E(T) = a^{1/b}\Gamma(1 + 1/b). \tag{7.15}$$

- T^b has an exponential distribution, with mean a.

- The hazard rate is $\lambda(t) = bt^{b-1}/a$.

If the shape parameter b is known, the scale parameter a can be estimated
by transforming to the exponential distribution. If $a = a(x)$ is modeled
using local likelihood, the Weibull model is a special case of the proportional
hazards model (7.7), with $\lambda_0(t) = bt^{b-1}$ and $\beta(x) = -\log(a(x))$.

Example 7.8. We return to the heart transplant dataset. The mean
survival time is modeled as a function of the patient's age at transplant
using the Weibull distribution. We estimate the scale parameter using a
local quadratic model and 80% nearest neighbor bandwidth and a global
constant model for the shape parameter b. By maximum likelihood, $\hat{b} =$
0.625, and the final fit is obtained as

```
> fit <- locfit(I((surv+0.5)^0.625)~age, cens=cens,
+    data=heart, alpha=0.8, family="gamma")
> y <- log(heart$surv+0.5)
> plotbyfactor(heart$age, heart$surv+0.5, heart$cens,
+    pch = c("0","+"), log = "y")
> plot(fit, add=T, tr = function(x)
+    exp(x/0.625)*gamma(1+1/0.625))
```

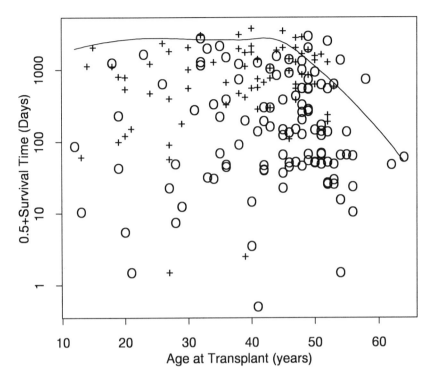

FIGURE 7.6. Heart transplant data and smooth using a censored local likelihood Weibull model.

The transformation `tr` provided to the `plot()` call arises from (7.15) and the log link $\theta = \log(a)$ used for the gamma family.

The fit in Figure 7.6 shows the same general pattern as the earlier fits in Figure 7.3. However, the overall level is raised, because the estimate is of log(mean survival time), rather than mean(log survival time).

7.4   Exercises

7.1 Consider the problem of censored local regression using a normal model. For the purposes of this exercise, suppose that $\sigma = 1$ is known.

a) The alternating (EM) algorithm proposed in Section 7.2 is performed globally. If instead the iterations are performed locally at each fitting point, show the resulting coefficients solve the equation

$$\sum_{i=1}^{n} w_i(x)A(x_i - x)A(x_i - x)^T \hat{a} = \sum_{i=1}^{n} w_i(x)A(x_i - x)\hat{Y}_i^* \quad (7.16)$$

where $\hat{Y}_i^*$ is obtained from (7.9) and (7.11), with μ_i replaced by $\langle \hat{a}, A(x_i - x) \rangle$.

b) An alternative is to use censored local likelihood with the Gaussian likelihood. Show the local likelihood equations for this case can also be expressed as (7.16). Thus, with known variance, the local likelihood and alternating algorithms coincide.

7.2 In the censored regression setting, Fan and Gijbels (1994) propose estimating Y_i^* using the transformation

$$\hat{Y}_i^* = \frac{\sum_{j:Y_j > Y_i, Y_j < c_j} W\left(\frac{x_j - x_i}{h}\right) Y_j}{\sum_{j:Y_j > Y_i, Y_j < c_j} W\left(\frac{x_j - x_i}{h}\right)}. \quad (7.17)$$

This is a local average of uncensored observations larger than Y_i. For definiteness, take $\hat{Y}_i^* = Y_i$ if the sums are empty.

a) Suppose responses T_i are exponentially distributed with mean μ and the censoring times c_i are exponentially distributed with mean τ; both T_i and c_i are independent of x_i. The observations are $Y_i = \min(T_i, c_i)$. Show

$$E(T_i | T_i \geq y) = y + \mu; \quad E(\hat{Y}_i^* | Y_i = y, Y_i = c_i) \leq y + \mu - \frac{\mu^2}{\mu + \tau}$$

(this would be equality if one ignores the problem of empty sums). Thus, the Fan/Gijbels method undercorrects in this case.

b) Propose a minor modification of (7.17) to fix the undercorrection in the exponential case. To what extent is the modified transformation nonparametric?

7.3 Suppose a random variable T has distribution function $F(t)$.

a) Show that the mean residual life is

$$E(T | T \geq y) - y = \frac{\int_y^\infty (1 - F(t))dt}{1 - F(y)}.$$

b) Suppose the hazard rate $\lambda(t)$ is a decreasing function of t. Show

$$1 - F(t) \leq \lambda(t) \int_t^\infty (1 - F(u))du$$

and hence

$$(1 - F(t))^2 \leq f(t) \int_t^\infty (1 - F(u)) du.$$

c) Show that if $\lambda(t)$ is a decreasing function of t, then the mean residual life is an increasing function of t. Also show for any $c > 0$,

$$E(T) \geq \frac{E(\min(T, c))}{P(T < c)}.$$

7.4 Suppose $T \sim N(0, \sigma^2)$ and $Y = \min(T, c)$ where c is a censoring time. Let

$$Y^* = Y I(Y < c) + \frac{\sigma \phi(Y/\sigma)}{1 - \Phi(Y/\sigma)} I(Y = c).$$

Show $E(YY^*) = \Phi(c) = P(Y < c)$. Given n observations (Y_i, c_i), show

$$\frac{\sum_{i=1}^n Y_i Y_i^*}{\sum_{i=1}^n I(Y_i < c_i)} \to \sigma^2$$

in probability, provided $\sum_{i=1}^n P(Y_i < c_i) \to \infty$. When σ is unknown, show the maximum likelihood estimate is the *solution* of the equation

$$\frac{\sum_{i=1}^n Y_i Y_i^*}{\sum_{i=1}^n I(Y_i < c_i)} = \sigma^2.$$

7.5 Let Y have a Poisson distribution with mean λ so that $P_\lambda(Y = y) = P_\lambda(y) = \lambda^y e^{-\lambda}/y!$. Show

$$\frac{d}{d\lambda} P_\lambda(Y = y) = P_\lambda(y - 1) - P_\lambda(y)$$

$$\frac{d}{d\lambda} P_\lambda(Y \geq y) = P_\lambda(y - 1).$$

Use these relations to show

$$\frac{d^2}{d\lambda^2} \log P_\lambda(Y \geq y)$$

$$= \frac{1}{P_\lambda(Y \geq y)^2} \sum_{j=y}^\infty (P_\lambda(y - 2) P_\lambda(j) - P_\lambda(y - 1) P_\lambda(j - 1)).$$

Hence show $\log P_\lambda(Y \geq y)$ is a concave function of λ. Derive a similar expression for $\frac{d^2}{d\theta^2} \log P_\lambda(Y \geq y)$ under the log link $\theta = \log(\lambda)$, and again show this is concave.

7.6 For the cricket batting dataset from Example 7.6, estimate and plot the hazard rate. Use a local log-linear model with smoothing parameter $\alpha = 0.4$ or smaller. Observe that the hazard rate displays an initial sharp decrease and then remains fairly constant.

8
Discrimination and Classification

In a classification problem, one has a set of multivariate observations from two (or more) populations. One then wishes to define a discriminant function that can effectively classify future observations into one of the populations. Classification problems have attracted an enormous amount of attention in recent years; reviews from a statistical perspective include Ripley (1994) and Langaas (1995). In this chapter we concentrate on statistical formulations of the classification problem and solutions using local likelihood methods.

Statistical work on classification begins with Fisher's (1936) linear discriminant analysis. One assumes that the populations give rise to normally distributed observations with different means but a common covariance matrix. Under this model, the optimal discriminant function can be shown to be a straight line.

More flexible classification rules can be obtained by relaxing the normality assumption and estimating the densities using local methods. The most common rules of this type are based on kernel density estimates (Van Ness and Simpson 1976; Kharin 1983; Murphy and Moran 1986) and nearest neighbor methods (Fix and Hodges 1951; Cover and Hart 1967).

An alternative formulation of the classification problem uses logistic regression (Day and Kerridge 1967; Anderson 1972; Efron 1975). In this approach, one attempts to estimate the probability that an observation comes from each population. An extensive study of local logistic regression for classification was provided by Deng and Moore (1996).

8.1   Discriminant Analysis

We consider the two-class classification problem, where observations arise from two populations, Π_1 and Π_2. Suppose one has a set of training observations (x_i, Y_i) where the responses Y_i are indicators of class membership; $Y_i = 1$ for observations from population Π_1, and $Y_i = 2$ for observations from Π_2. One then wishes to define a discriminant function that can effectively classify a future observation into one of the two populations. A decision rule $\delta(x)$ maps the sample space to $\{1, 2\}$. An observation $X \in \Pi_1$ is correctly classified if $\delta(X) = 1$ and misclassified if $\delta(X) = 2$.

Suppose that for observations from Π_1, x_i has a density $f_1(x)$, and for observations from Π_2, x_i has a density $f_2(x)$. Suppose also that the populations have prior probabilities

$$
\begin{aligned}
p_1 &= P(x_i \in \Pi_1) \\
p_2 &= P(x_i \in \Pi_2)
\end{aligned}
$$

with $p_1 + p_2 = 1$. Applying Bayes' theorem, the posterior probability that an observation x_i comes from the population Π_1 is

$$
P(x_i \in \Pi_1 | x_i = x) = \frac{p_1 f_1(x)}{p_1 f_1(x) + p_2 f_2(x)} = \left(1 + \frac{p_2 f_2(x)}{p_1 f_1(x)}\right)^{-1}. \tag{8.1}
$$

We also introduce a loss function. Suppose we have an observation $X = x$ but the corresponding population is unknown. Suppose the observation is classified as $\delta(x) = 2$. If the observation is from Π_2, the observation is correctly classified and the cost is 0. If the observation is from Π_1, the object is misclassified, and the cost is c_1. The expected loss is then c_1 times the probability that the observation has been misclassified:

$$
L(x, 2) = c_1 P(X \in \Pi_1 | X = x) = \frac{c_1 p_1 f_1(x)}{p_1 f_1(x) + p_2 f_2(x)}. \tag{8.2}
$$

Likewise, if misclassifying an observation from Π_2 as coming from Π_1 has cost c_2, the expected cost of $\delta(x) = 1$ is

$$
L(x, 1) = c_2 P(X \in \Pi_2 | X = x) = \frac{c_2 p_2 f_2(x)}{p_1 f_1(x) + p_2 f_2(x)}. \tag{8.3}
$$

An optimal Bayes rule $\delta_B(x)$ chooses $\delta_B(x) = 1$ for values of x with $L(x, 1) < L(x, 2)$, and $\delta_B(x) = 2$ otherwise. Explicitly,

$$
\delta_B(x) = \begin{cases} 1 & c_1 p_1 f_1(x) > c_2 p_2 f_2(x) \\ 2 & c_1 p_1 f_1(x) \le c_2 p_2 f_2(x) \end{cases} \tag{8.4}
$$

or equivalently,

$$
\delta_B(x) = \begin{cases} 1 & P(X \in \Pi_2 | X = x) < \frac{c_1}{c_1 + c_2} \\ 2 & P(X \in \Pi_2 | X = x) \ge \frac{c_1}{c_1 + c_2} \end{cases} . \tag{8.5}
$$

The global expected loss of the optimal rule $\delta_B(x)$ is obtained by integrating the pointwise loss:

$$c_1 P(X \in \Pi_1, \delta_B(X) = 2) + c_2 P(X \in \Pi_2, \delta_B(X) = 1)$$

$$= c_1 \int_{\{x:\delta_B(x)=2\}} p_1 f_1(x) dx + c_2 \int_{\{x:\delta_B(x)=1\}} p_2 f_2(x) dx$$

$$= \int \min(c_1 p_1 f_1(x), c_2 p_2 f_2(x)) dx. \qquad (8.6)$$

The optimal rule $\delta_B(x)$ cannot be used directly since $f_1(x)$ and $f_2(x)$ are unknown. Instead, we have to build data-based classifiers that approximate, as closely as possible, this optimal rule. To estimate the optimal discriminant rule, (8.4) and (8.5) suggest two distinct approaches. First, we can estimate $P(X \in \Pi_2 | X = x)$ directly using local logistic regression. Alternatively, we can estimate $f_1(x)$ and $f_2(x)$ using density estimation.

8.2 Classification with LOCFIT

To illustrate the classification methods, we use a simulated example with two classes Π_1 and Π_2, and two predictor variables. Under Π_1, we generate $x_{1,i} \sim N(0,1)$ and $x_{2,i} = (x_{1,i}^2 - 2 + z_i)/3$, where $z_i \sim N(0,1)$. Under the class Π_2, $x_{1,i} \sim N(0,1)$ and $x_{2,i} = (2 - x_{1,i}^2 + z_i)/3$. The two classes are equally likely, and misclassification costs are equal, $c_1 = c_2 = 1$. Under this model, the optimal classifier $\delta_B(x)$ has a checkerboard pattern, with a horizontal boundary at $x_2 = 0$, and vertical boundaries at $x_1 = \pm\sqrt{2}$. By numerical integration, the global expected cost (8.6) is 0.1094, so that about 11% of the observations should be misclassified by $\delta_B(x)$.

A training sample with 200 observations was generated:

```
> y <- sample(c(1,2),200,replace=T)
> x1 <- rnorm(200)
> x2 <- (x1*x1-2+rnorm(200))/3*(1-2*y)
> cltrain <- data.frame(x1=x1,x2=x2,y=y)
```

A test set `cltest` was generated similarly, also with 200 observations. The training set contains 93 observations from Π_1 and 107 from Π_2. The test set had 109 observations from Π_1 and 91 from Π_2. Table 8.1 summarizes the performance of $\delta_B(x)$ on the test and training samples; this forms a useful basis for comparing data-based rules later. There are 27 misclassifications (13.5%) on the training sample and 17 (8.5%) on the test sample.

		Train		Test	
		1	2	1	2
Classified	1	81	15	99	7
as	2	12	92	10	84

TABLE 8.1. Classification rates for $\delta_B(x)$ on a simulated example, with training and test samples.

8.2.1 Logistic Regression

Local logistic regression can then be applied directly to estimate the posterior probability (8.1) of class membership and hence the discriminant rule.

Example 8.1. We fit a local logistic regression to the training dataset generated earlier;

```
> fit <- locfit(I(y==2)~x1+x2, data=cltrain, scale=0)
```

Note the binomial family is automatic when the response is logical. The fit is then ploted, with a single contour at the 0.5 level:

```
> plot(fit, v=0.5)
> text(cltrain$x1, cltrain$x2, cltrain$y, cex=0.7)
```

Figure 8.1 shows the data and discriminant boundary. Although the sharp corners of the optimal boundary have been rounded off, the estimated boundary generally follows the same pattern as the optimal boundary. The classification table shows a total of 27 misclassifications:

```
> table(fitted(fit)>0.5, class.train$y)
        1  2
FALSE 80 14
 TRUE 13 93
```

The fitted object can be used to classify the test sample, simply by evaluating the fit at test points. The classification rule is evaluated entirely from the fitted object using the predict() function; it does not use the original data. This speeds up computation when the training sample is large.

Example 8.2. We compute the predicted values for the fit computed in Example 8.1, at the test dataset:

```
> table(predict(fit,cltest)>0.5, cltest$y)
        1  2
FALSE 98  9
 TRUE 11 82
```

Here, we have misclassified 20 observations, compared to 17 using $\delta_B(x)$. As expected, the data-based rule is slightly worse than the optimal rule.

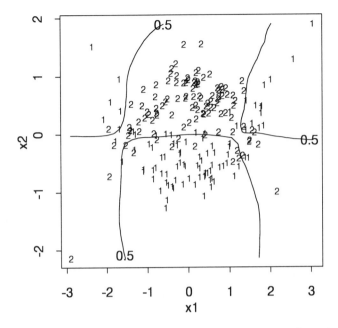

FIGURE 8.1. Classification boundary for a simulated example, using logistic regression.

8.2.2 Density Estimation

The second approach to discrimination uses density estimation. This takes a little more work to implement, since separate fits must be computed for each class. We actually use Poisson process rate estimation (family="rate"). As shown in Section 5.1.2, this has the effect of multiplying the estimate by the sample size. This is desirable here since the decision rule (8.4) involves the prior probabilities p_1 and p_2 and these are estimated as $n_1/(n_1 + n_2)$ and $n_2/(n_1 + n_2)$ respectively. Implicit here is the assumption that the sampling frequencies in the training sample reflect the general population.

Example 8.3. For each of the two classes in Example 8.1, we compute density estimates:

```
> fit2 <- locfit(~x1+x2, data=class.train, subset=y==2,
+   family="rate", scale=0)
> fit1 <- locfit(~x1+x2, data=class.train, subset=y==1,
+   family="rate", scale=0)
```

To tabulate the performance of the discriminant rule on the training sample, the two fits are predicted at *all* the sample points:

```
> id <- function(x)x
```

```
> fiy1 <- predict(fit1, class.train, tr=id)
> fiy2 <- predict(fit2, class.train, tr=id)
> table(fiy2-fiy1>0, class.train$y)
        1  2
FALSE 81 14
 TRUE 12 93
```

The predicted values are computed with the identity inverse transformation, so the returned values are estimates of the logarithm of the event rate. This avoids numerical division-by-0 errors in sparse regions. We now have 26 misclassifications on the training sample; slightly better than $\delta_B(x)$ in Table 8.1. This is to be expected; the fitted rule is tuned to the training sample at hand, whereas the optimal rule $\delta_B(x)$ is tuned to the population.

Plotting the discriminant region also takes a little care. Since the two fits are not computed at the same points, we cannot directly subtract the fits. Rather, we must predict each fit on the same grid of points, and then subtract the predictions:[1]

```
> pr <- lfmarg(c(-3,-2.2,3,2),c(50,50))
> plot(preplot(fit2,pr,tr=id)-preplot(fit1,pr,tr=id), v=0)
> text(class.test$x1, class.test$x2, class.test$y, cex=0.7)
```

The lfmarg() function is used to compute a grid of points over an appropriate prediction region; here, $[-3, 3] \times [-2.2, 2]$, with 50 points per side. The discriminant boundary is shown in Figure 8.2. In this case the estimated boundary is quite unlike the optimal boundary. But the differences are largely in regions where there is very little data, so this has little effect on the misclassification rates.

To classify the test dataset, we need to evaluate the two fits at the points in the test dataset:

```
> pry1 <- predict(fit1, class.test, tr = id)
> pry2 <- predict(fit2, class.test, tr = id)
> table(pry2-pry1>0, class.test$y)
         1  2
FALSE 100  7
 TRUE   9 84
```

This produces only 16 misclassifications, beating the optimal rule $\delta_B(x)$ on the test set. This can only happen by chance, since the expected loss for the data-based rule must be larger than for the optimal rule.

---

[1] This will only work in S version 4 and higher.

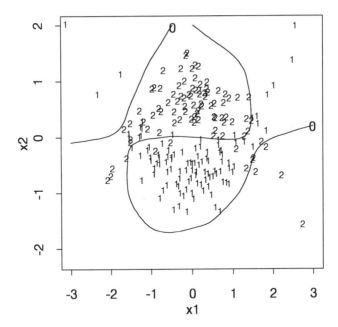

FIGURE 8.2. Classification boundary for a simulated example, using density estimation.

8.3   Model Selection for Classification

The usual model issues: choice of local polynomial order and bandwidth selection, arise in the classification problem. However, they manifest themselves in different ways, and choices that result in good visual estimates of the densities and probability functions need not be the best choices for classification.

Variable selection is also important in classification problems. Adding new variables to the model reduces the optimal Bayes risk (8.6) (see Exercise 8.2). But in practice, adding variables leads to a more difficult estimation problem, and it may be best to use only a subset of the available variables.

The classification problem is ideally suited to cross validation. We ask the question: "how would each observation be classified, when the classification rule is defined from the remaining $n-1$ observations?". The process is best studied by examples.

Example 8.4. Fisher's iris dataset (Fisher 1936) consists of four measurements on iris flowers: petal width, petal length, sepal width and sepal length. We consider classification of the Virginica and Versicolor species (a third species, Setosa, is in the original dataset but is not used here).

Each model with one or two variables is fitted local linear logistic regression and a 70% nearest neighbor smoothing parameter. Direct leave-one-out cross validation is performed at each data point (this is feasible with only 100 observations), so each observation is being classified by the remaining 99 observations. The misclassification rate is then computed. A typical call is

```
> fit <- locfit(I(species=="virginica")~sepal.len, data=iris,
+   deg=1,ev="crossval")
> table(fitted(fit)>0.5, iris$species)
        versicolor virginica
FALSE          34         11
 TRUE          16         39
```

In this case, $27 = 16 + 11$ misclassifications result. For multiple variables, the scale=0 argument was added.

		No. of Misclassifications	
Variables	Versicolor	Virginica	Total
Sepal length	16	11	27
Sepal width	23	21	44
Petal length	4	3	7
Petal width	2	4	6
Petal width, sepal length	2	5	7
Petal width, sepal width	4	6	10
Petal width, petal length	3	3	6

TABLE 8.2. Variable selection by cross validation for the iris data.

Table 8.2 shows the results for all one and two variable models. Clearly, sepal width and sepal length have high misclassification rates, while petal length and petal width have lowest misclassification rates. On this basis, petal width is selected as the first variable in the model. Adding a second variable provides no improvement. Figure 8.3 shows the classification boundary when petal width and petal length are included in the model:

```
> fit <- locfit(I(species=="virginica")~petal.wid+petal.len,
+   data=iris, scale=0, deg=1)
> plotbyfactor(petal.wid, petal.len, species, data=iris,
+   pch=c("O","+"), lg = c(1, 6.8))
> plot(fit, v=0.5, add=T)
```

What about the bandwidth and degree? As we have seen, these can be crucial in effective modeling of regression surfaces, particularly in problems where there is large curvature. But in classification problems, bandwidth and degree often have less effect. The reason is that extremities of the

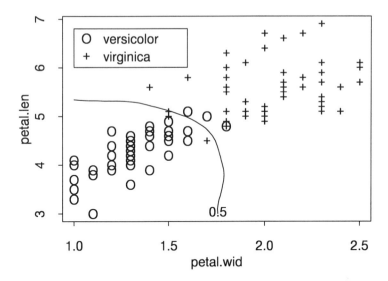

FIGURE 8.3. Classification boundary for the iris dataset, based on petal width and petal length.

fitted surface are less important. For example, modeling the densities of the individual species in Figure 8.3 would need a local quadratic model to capture peaks for the two species. But in classification, we only need good estimates in the small region where the two species mix. Outside this region, it is obvious what the classification rule should be, and even poor density estimates will get this right.

Example 8.5. To see the effect of changing the smoothing parameter on the iris data, we fit the bivariate `petal.len+petal.wid` model and vary the nearest neighbor span from 0.2 to 0.9. Table 8.3 shows that the smoothing parameter is having very little effect on the cross validated error rate. This pattern is commonly observed in classification problems; for many real datasets, selection of variables and data transformation are far more important than bandwidth selection.

α	0.2	0.3	0.4	0.5	0.6	0.7	0.8	0.9
Misclass. rate	5	5	7	6	6	6	6	6

TABLE 8.3. Effect of changing the smoothing parameter α on the cross validated error rate for the iris data.

How do these results compare with other approaches to classification? In Table 3 of Friedman (1994) the results of eight different classifiers are reported, including CART, k-nearest neighbor and proposed scythe and machete methods. The results ranged from 3 to 11 misclassifications, with five of the eight classifiers having 5 or 6 misclassifications. Clearly the results obtained by logistic regression are comparable.

So far we have performed leave-one-out cross validation directly. This is reasonable for the iris dataset with just 100 points; with larger sample sizes it is less reasonable. An alternative is to use the approximate cross validation based on influence functions developed in Section 4.3.3 for local likelihood models. The implementation is quite simple; for example,

```
> fit <- locfit(I(species=="virginica")~sepal.len,
+    data=iris, deg=1)
> table(fitted(fit,cv=T)>0.5, iris$species)
      versicolor virginica
FALSE        34        11
 TRUE        16        39
```

Compared to the direct fit, we drop the ev="crossval" argument from the call to locfit() and add cv=T to the call to fitted(). In this case, the result is identical to the direct cross validation. For the seven models reported in Table 8.2, the approximate cross validation produced identical results in five cases; the exceptions being petal.wid+sepal.len (6 misclassifications) and petal.wid+sepal.wid (8 misclassifications).

8.4 Multiple Classes

The classification problem can be extended to multiple classes. Suppose K populations $\Pi_1, \ldots, \Pi_K$ have prior probabilities $p_1, \ldots, p_K$ and densities $f_1(x), \ldots, f_K(x)$. Similarly to (8.1), the posterior probability of class i is

$$P(x_i \in \Pi_j | x_i = x) = \frac{p_j f_j(x)}{p_1 f_1(x) + \ldots + p_K f_K(x)}.$$

Assuming a 0-1 loss function, the optimal decision rule selects the class with maximum posterior probability,

$$\delta(x) = \text{argmax}_{1 \leq j \leq K} p_j f_j(x).$$

As in the two-class problem, we can estimate the posterior probabilities using local logistic regression or the class densities $p_i f_i(x)$ using density estimation.

Example 8.6. We use the chemical and overt diabetes dataset from Reaven and Miller (1979) to classify type of diabetes from several predictor variables. Some inspection shows that one variable, 'Glucose Area', provides good discrimination between the classes; fitting local linear logistic regression produces a cross validated error rate of $6/145 = 4.1\%$. Adding

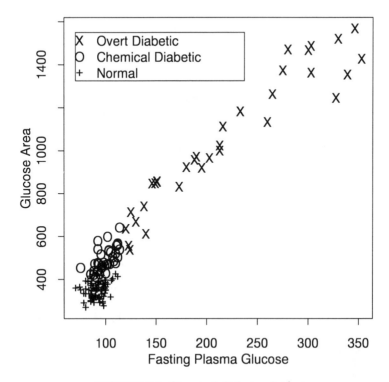

FIGURE 8.4. Chemical diabetes dataset.

a second variable, 'Fasting Plasma Glucose', produces nearly perfect discrimination; see Figure 8.4.

For each of the three classes, we compute the cross validated local logistic regressions; for example,

```
> fit1 <- locfit(I(cc=="Overt Diabetic")~fpg+ga,
+    data=chemdiab, scale=0, deg=1, ev="crossval")
> p1 <- fitted(fit1)
```

Predictions p2 and p3 are computed for the other responses ("Chemical Diabetic" and "Normal") respectively. We then compute and tabulate the classifications:

```
> z <- (p1>pmax(p2,p3)) + 2*(p2>pmax(p1,p3))
+    + 3*(p3>pmax(p1,p2))
> table(chemdiab$cc, z)
                  1  2  3
  Overt Diabetic 33  0  0
Chemical Diabetic  0 35  1
          Normal  0  1 75
```

This two variable model has a cross validated error rate of $2/145 = 1.4\%$; this significantly beats the eight methods in table 3 of Friedman (1994).

Example 8.7. The kangaroo skull dataset (Andrews and Herzberg, 1985 chapter 53) consists of 18 skull measurements on 148 kangaroos from three species. Attention is restricted to the 101 skulls with a complete set of measurements. Using local linear logistic regression, the first variable entered is crest width, resulting in a cross validated error rate of $45/101 = 44.6\%$. The second variable entered is ascending ramus height, with error rate $31/101 = 30.7\%$. An analysis of the misclassifications (Table 8.4) shows the species giganteus and fuliginosus are being successfully distinguished, while melanops is proving difficult to identify.

	Classified as		
Species	Gig.	Mel.	Ful.
Giganteus	32	3	2
Melanops	13	5	10
Fuliginosus	2	1	33

TABLE 8.4. Classification rates for the kangaroo dataset.

Figure 8.5 displays the two selected variables. This confirms what we observe from Table 8.4: The giganteus and fuliginosus species are fairly well separated from each other, but melanops is difficult to distinguish. Results in table 3 of Friedman (1994) show misclassification rates ranging from 19.8 to 48.0%; the performance of the bivariate model is average. Adding additional variables doesn't help much, and one begins to get numerical problems because local logistic regression with three variables and just 101 observations is difficult.

Variable selection for the kangaroo skulls was only moderately successful. The reason is that the eighteen variables each contribute a small amount of information. An alternative approach is to look not just at marginal variables, but at low dimensional projections of the data. One approach is to look at the projections produced by linear discriminant analysis. Suppose, for the moment, that observations from each species have a multivariate normal distribution, with a common covariance matrix but different means. That is, we have the model

$$X_{i,j} = \mu_i + Z_{i,j}; i = 1, 2, 3; j = 1, \ldots, n_j.$$

Here, $X_{i,j}$ is the vector of measurements for the jth individual from the ith population, and μ_i is the mean for the ith population. The $Z_{i,j}$ are independent multivariate normal random vectors with mean 0 and covariance Σ. Let $\Sigma = \mathbf{P}\mathbf{D}\mathbf{P}^T$ be the eigen decomposition of Σ. Then consider

$$Y_{i,j} = \mathbf{D}^{-1/2}\mathbf{P}^T X_{i,j} = \mathbf{D}^{-1/2}\mathbf{P}^T \mu_i + \mathbf{D}^{-1/2}\mathbf{P}^T Z_{i,j}.$$

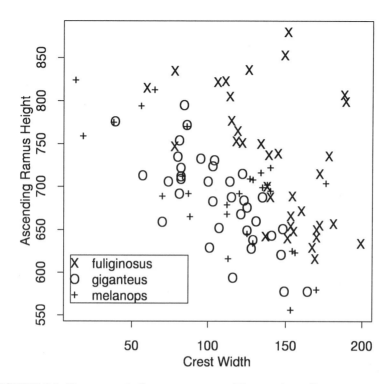

FIGURE 8.5. Kangaroo skull measurements: The two best discrimination variables.

Under the normal model, all the discriminatory power is in the projection of $Y_{i,j}$ onto the plane containing $\mathbf{D}^{-1/2}\mathbf{P}^T(\mu_1, \mu_2, \mu_3)$; directions orthogonal to this plane are irrelevant. Of course, we probably don't believe the normal model, but this canonical transformation provides two variables that can be used for a fit.

Example 8.8. We apply canonical rotations to the kangaroo dataset. The resulting cross validated classification rates are shown in Table 8.5, and the classification boundary in Figure 8.6. Here, we have been able to mostly separate the melanops species, and the cross validated error rate is now $13/101 = 12.9\%$. This significantly beats the model based on marginal variables and comfortably beats the best of the methods considered in Friedman (1994).

	Classified as		
Species	Gig.	Mel.	Ful.
Giganteus	32	5	0
Melanops	7	20	1
Fuliginosus	0	0	36

TABLE 8.5. Cross validation classification rates for the kangaroo data, based on canonical variables.

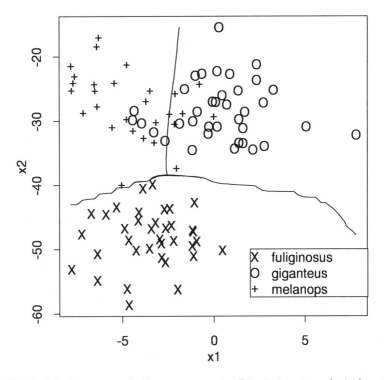

FIGURE 8.6. Kangaroo skull measurements: Discriminant analysis based on canonical variables.

8.5   More on Misclassification Rates

In this section we look further at the error rates of the classification rules for the two population case. We take two approaches: a pointwise approach (based on the work of Friedman (1997)) and a global approach. The results are used to compare the density estimation and logistic regression approaches.

8.5.1 Pointwise Misclassification

A data-based decision rule $\delta(x)$ can be considered (for fixed x) a random variable, dependent on the training sample. Conditioned on $\delta(x)$, the expected loss is given by (8.3) or (8.2) as appropriate. Taking expectations with respect to the distribution of the training sample gives the expected loss:

$$P(\delta(x) = 1)c_2 P(X \in \Pi_2 | X = x) + P(\delta(x) = 2)c_1 P(X \in \Pi_1 | X = x). \quad (8.7)$$

For the optimal rule, $\delta_B(x)$ is deterministic, and (except possibly on the classification boundary) $P(\delta_B(x) = 1)$ will be 0 or 1. Suppose, for convenience, that we are at a point x for which $\delta_B(x) = 1$. Then the expected loss of $\delta_B(x)$ is $c_2 P(X \in \Pi_2 | X = x)$, and the expected loss (8.7) of a data-based rule $\delta(x)$ can be written (using $P(\delta(x) = 1) = 1 - P(\delta(x) = 2)$)

$$c_2 P(X \in \Pi_2 | X = x)$$
$$+ P(\delta(x) = 2)(c_1 P(X \in \Pi_1 | X = x) - c_2 P(X \in \Pi_2 | X = x)). \quad (8.8)$$

The first term of (8.8) is the expected loss of the optimal rule $\delta_B(x)$. The second term (which is always positive) represents the increase in expected loss due to the use of the data-based decision rule $\delta(x)$. The only component that depends on the choice of decision rule is $P(\delta(x) = 2)$, or more generally, $P(\delta(x) \neq \delta_B(x))$. Thus, a good decision rule will make this probability small.

For many classification rules - including those based on local likelihood methods discussed earlier - the decision rule $\delta(x)$ is based on an estimate $\hat{p}(x)$ of the posterior probability $P(X \in \Pi_1 | X = x)$, and $\hat{p}(x)$ will have an asymptotically normal distribution. In this case, $P(\delta(x) \neq \delta_B(x))$ is simply a tail probability for $\hat{p}(x)$.

Suppose the approximating normal distribution has mean $p^*(x)$ and variance $\sigma(x)$. These can be derived using (4.19) when $\hat{p}(x)$ is constructed using local logistic regression or (5.23) for density estimates. Let λ be the critical value for classification; $\lambda = c_1/(c_1 + c_2)$ by (8.5). For points x with $\delta_B(x) = 1$, $p(x) \geq \lambda$. Thus, $P(\delta(x) \neq \delta_B(x)) = P(\hat{p}(x) < \lambda)$. Using the normal approximation,

$$P(\hat{p}(x) < \lambda) \approx \frac{1}{\sigma(x)} \int_{-\infty}^{\lambda} \phi\left(\frac{v - p^*(x)}{\sigma(x)}\right) dv$$
$$= \Phi\left(\frac{\lambda - p^*(x)}{\sigma(x)}\right).$$

In general,

$$P(\delta(x) \neq \delta_B(x)) \approx \Phi\left(\mathrm{sgn}(p(x) - \lambda)\frac{\lambda - p^*(x)}{\sigma(x)}\right).$$

This error approximation has some interesting properties. Usually, we expect $p(x) - \lambda$ and $p^*(x) - \lambda$ to have the same sign. In this case, it is advantageous for the estimate to have small variance; $P(\delta(x) \neq \delta_B(x)) \to 0$ as $\sigma(x) \to 0$. But if $\hat{p}(x)$ is heavily biased or x is near the optimal classification boundary, $p(x) - \lambda$ and $p^*(x) - \lambda$ may have opposite signs. In this case, small variance is a penalty: $P(\delta(x) \neq \delta_B(x)) \to 1$ as $\sigma(x) \to 0$.

The conclusion here is that small classification error requires ensuring, as far as possible, that $p(x) - \lambda$ and $p^*(x) - \lambda$ have the same sign. The usual goodness-of-fit measures, such as squared error, are of less relevance.

8.5.2   Global Misclassification

The global (averaged over x) misclassification rate is

$$\int_{\{x:\delta(x)=1\}} c_2 p_2 f_2(x)dx + \int_{\{x:\delta(x)=2\}} c_1 p_1 f_1(x)dx.$$

For the optimal rule $\delta_B(x)$, this becomes

$$\int \min(c_1 p_1 f_1(x), c_2 p_2 f_2(x))dx.$$

The rate for the data-based rule can thus be written

$$\int \min(c_1 p_1 f_1(x), c_2 p_2 f_2(x))dx + \int_{\delta(x) \neq \delta_B(x)} |c_1 p_1 f_1(x) - c_2 p_2 f_2(x)|dx.$$

We restrict attention to the case where x is one dimensional. In this case, there will (usually) be a discrete set of boundary points where $c_1 p_1 f_1(x) = c_2 p_2 f_2(x)$ and $\delta_B(x)$ switches from 1 to 2. Assuming our density estimates $\hat{f}_1(x)$ and $\hat{f}_2(x)$ are uniformly consistent, then most misclassifications will occur in the neighborhood of boundary points.

This point is clarified in Figure 8.7. Here, a dataset has been simulated with $x_i \sim N(0,1)$ and $P(Y_i = 1|x_i = x) = e^x/(1 + e^x)$. Clearly, the optimal classification rule (assuming equal costs) is $\delta_B(x) = 0$ for $x < 0$, and $\delta_B(x) = 1$ for $x \geq 0$. The estimated probability, using a local quadratic model, is shown by the solid line in Figure 8.7. Especially in the tails, this isn't a very good estimate. But mostly this doesn't matter; only in the small shaded region does the estimated rule differ from the optimal rule. Thus, to determine the effect of estimating the classification rule, we need only focus on this region.

For simplicity, suppose there is just a single boundary point x^*, and the estimated decision rule results in an estimated boundary point $\hat{x}^*$. In a neighborhood of x^*, we use linear expansions $f_1(x) \approx f_1(x^*) + (x - x^*)f_1'(x^*)$. Then

$$\int_{\delta(x) \neq \delta_B(x)} |c_1 p_1 f_1(x) - c_2 p_2 f_2(x)|dx \qquad (8.9)$$

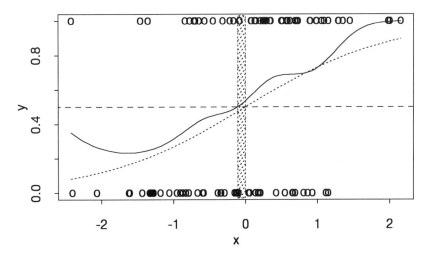

FIGURE 8.7. Estimating the classification boundary. For a simulated dataset, the true class probability is indicated by the short dashed line and the estimated class probability by the solid line. The estimated classification rule differs from the optimal rule only in the vertical bar.

$$= \int_{[x^*,\hat{x}^*]} |c_1 p_1 f_1(x) - c_2 p_2 f_2(x))| dx$$

$$\approx \int_{[x^*,\hat{x}^*]} |x - x^*| dx \cdot |c_1 p_1 f_1'(x^*) - c_2 p_2 f_2'(x^*)|$$

$$= \frac{1}{2}(\hat{x}^* - x^*)^2 \cdot |c_1 p_1 f_1'(x^*) - c_2 p_2 f_2'(x^*)|. \qquad (8.10)$$

Thus, the global misclassification rate requires a good estimate of the boundary point x^*, as measured by squared error loss $(\hat{x}^* - x^*)^2$. Since $\hat{x}^*$ is the solution of $\hat{p}(x) = p(x^*)$, another linear expansion gives

$$\hat{p}(\hat{x}^*) \approx \hat{p}(x^*) + (\hat{x}^* - x^*)\hat{p}'(x^*)$$

$$\hat{x}^* - x^* \approx -\frac{p(\hat{x}^*) - p(x^*)}{\hat{p}'(x^*)}.$$

Substituting into (8.10), the classification error is largely dependent on the squared estimation error at the boundary point, $(p(\hat{x}^*) - p(x^*))^2$. This is quite different from the pointwise misclassification results obtained previously. Note that variability in $\hat{p}'(x^*)$ can be ignored asymptotically, provided $p'(x^*)$ is nonzero.

8.6   Exercises

8.1 A nearest neighbor classifier places an observation $X = x$ in the same class as the nearest observation in the training set. In the two-class setting, suppose the densities $f_0(x)$ and $f_1(x)$ are continuous, and let $n \to \infty$.

a) For fixed x, show

$$P(\delta(x) = 0) \to \frac{p_0 f_0(x)}{p_0 f_0(x) + p_1 f_1(x)}.$$

b) Show the probability of misclassification, conditional on $X = x$, converges to

$$2\frac{p_0 p_1 f_0(x) f_1(x)}{(p_0 f_0(x) + p_1 f_1(x))^2}. \tag{8.11}$$

c) Show the error rate (8.11) is less than twice the corresponding quantity for the optimal rule $\delta_B(x)$. Generalize this to k populations (these results were first proven by Cover (1968)).

8.2 Consider a classification problem with two predictors $X = (X_1, X_2)$. Under Π_1 the observations have density $f_1(x_1, x_2)$, and under Π_2, the density is $f_2(x_1, x_2)$. Let $g_1(x_1)$ and $g_2(x_1)$ be the marginal densities of X_1 under the two populations. Show

$$\int_{x_1} \int_{x_2} \min(f_1, f_2) dx_1 dx_2 \le \int_{x_1} \min(g_1, g_2) dx_1.$$

Thus show that adding the variable X_2 to the model reduces the optimal Bayes loss.

Hint. Write f_1 and f_2 in terms of g_1, g_2 and the conditional densities of X_2 given X_1. Then, note that $\min(h_1 g_1, h_2 g_2) \le h_1 g_1$ on the set $\{g_1 \le g_2\}$ and $\min(h_1 g_1, h_2 g_2) \le h_2 g_2$ otherwise.

8.3 Consider the training sample $(x_i, Y_i); i = 1, \ldots, n$, with x_i in the predictor space and $Y_i \in \{0, 1\}$ the indicator for class membership.

a) Use local constant logistic regression to estimate the class probabilities. Show the estimate of $P(X \in \Pi_1 | X = x)$ is

$$\hat{P}(X \in \Pi_1 | X = x) = \frac{\sum_{i=1}^{n} W((x_i - x)/h) I(Y_i = 1)}{\sum_{i=1}^{n} W((x_i - x)/h)}. \tag{8.12}$$

b) Use local constant density estimation to estimate $f_1(x)$ and $f_2(x)$. Use the same bandwidth for each estimate. Show

$$\hat{f}_1(x) = \frac{1}{n_1 h} \sum_{i=1}^{n} W\left(\frac{x_i - x}{h}\right) I(Y_i = 1)$$

$$\hat{f}_2(x) \quad = \quad \frac{1}{n_2 h} \sum_{i=1}^{n} W\left(\frac{x_i - x}{h}\right) I(Y_i = 2).$$

Compute an estimate of the probability $P(X \in \Pi_1 | X = x)$ from the density estimates (use $\hat{p}_1 = n_1/(n_1 + n_2)$). Compare with (8.12).

8.4 The urine crystal dataset (Andrews and Herzberg, (1985), chapter 44) contains six chemical measurements and an indicator for the presence of calcium crystals. Use the 77 observations with no missing values.

  a) Using local linear logistic regression at the default bandwidths, show the best single predictor is calcium concentration, with a cross validated error rate of $19/77 = 24.7\%$. Setting scale=0, show that the second variable entered is urea concentration, with a cross validated error rate of $17/77 = 22.1\%$.

  b) Using calcium and urea concentrations as predictors, experiment with changing some or all of the bandwidths, degree of fit and scale parameters. Obtain a cross validated error rate of $14/77 = 18.2\%$ or better. Compare these results with the results of Friedman (1994).

8.5 (Research Problem). The approximation of global misclassification rates in Section 8.5.2 applies only when a single predictor variable is used for classification. Derive similar results for two (or higher) dimensions, when the classification boundary is a curve (or surface). Extend the results to problems with three or more classes.

9
Variance Estimation and Goodness of Fit

In this chapter we study inferential issues for local regression. Section 9.1 studies variance estimation, including distributional approximations and methods for handling nonhomogeneous variance. This is applied to goodness of fit testing using generalizations of the F test. Section 9.2 discusses construction of confidence intervals and bands around local regression estimates.

9.1 Variance Estimation

For many diagnostic purposes, such as the CP statistic introduced in Section 2.4.1, confidence bands and the goodness of fit tests introduced later in this chapter, it is necessary to estimate the error variance σ^2. Estimates of σ^2 are also of interest in their own right, since σ^2 represents the amount of variation in the data that cannot be explained by the predictor variables. This should be contrasted with the widely reported squared correlation coefficient R^2, which is the fraction of variation explained by the predictor variables. But R^2 often has little meaning, since it is dictated as much by the choice of the data points x_i as by the strength of the relation between variables. One can always increase R^2, simply by repeating an experiment but spreading the x_i points out more. But as long as the error variance σ^2 remains constant, variance estimates estimate the same quantity for both experiments.

In Section 2.3.2 we considered the residual variance estimate

$$\hat{\sigma}^2 = \frac{1}{n - 2\nu_1 + \nu_2} \sum_{i=1}^{n} (Y_i - \hat{\mu}(x_i))^2. \qquad (9.1)$$

The expected value follows from (2.32), assuming the errors ϵ_i are independent with $E(\epsilon_i^2) = \sigma^2$:

$$E(\hat{\sigma}^2) = \sigma^2 + \frac{1}{n - 2\nu_1 + \nu_2} \sum_{i=1}^{n} \text{bias}(\hat{\mu}(x_i))^2.$$

In particular, $\hat{\sigma}^2$ is unbiased when the estimate $\hat{\mu}(x)$ is unbiased. This generally holds only for a small class of mean functions, although with small bandwidths it may be reasonable to assume the bias is negligible.

What is the distribution of $\hat{\sigma}^2$? The residual sum of squares can be written as a quadratic form,

$$\sum_{i=1}^{n} (Y_i - \hat{\mu}(x_i))^2 = \|(\mathbf{I} - \mathbf{L})Y\|^2 = Y^T \mathbf{\Lambda} Y \qquad (9.2)$$

where $\mathbf{L}$ is the hat matrix and $\mathbf{\Lambda} = (\mathbf{I} - \mathbf{L})^T (\mathbf{I} - \mathbf{L})$. Thus,

$$\hat{\sigma}^2 = \frac{1}{\text{tr}(\mathbf{\Lambda})} Y^T \mathbf{\Lambda} Y. \qquad (9.3)$$

The distribution can be found through an eigenvalue decomposition of the matrix $\mathbf{\Lambda}$. In particular, if the errors ϵ_i are normally distributed and $\hat{\sigma}^2$ is unbiased (i.e., $\mu^T \mathbf{\Lambda} \mu = 0$), the distribution of the quadratic form is

$$Y^T \mathbf{\Lambda} Y \overset{\mathcal{L}}{=} \sigma^2 \sum_{j=1}^{n} \lambda_j Z_j \qquad (9.4)$$

where λ_j are the eigenvalues of $\mathbf{\Lambda}$ and Z_j are independent χ_1^2 random variables.

For a parametric regression model, the λ_j are all 0 or 1 (i.e., $\mathbf{\Lambda}$ is idempotent), and $Y^T \mathbf{\Lambda} Y / \sigma^2$ has a χ^2 distribution. For local regression variance estimates, this simplification no longer holds. Numerical methods for finding the distribution of quadratic forms have been discussed in Imhof (1961) and Davies (1980).

Since $\mathbf{\Lambda}$ is an $n \times n$ matrix, finding the eigenvalues and computing the exact distribution is expensive for large n. But simple approximations can be derived using the mean and variance of $\hat{\sigma}^2$. Since $E(Z_i) = 1$ and $\text{var}(Z_i) = 2$, (9.3) and (9.4) yield

$$E(\hat{\sigma}^2) = \frac{\sigma^2}{\text{tr}(\mathbf{\Lambda})} \sum_{i=1}^{n} \lambda_i = \sigma^2 \text{tr}(\mathbf{\Lambda})$$

$$\text{var}(\hat{\sigma}^2) = 2 \frac{\sigma^4}{\text{tr}(\mathbf{\Lambda})^2} \sum_{i=1}^{n} \lambda_i^2 = 2\sigma^4 \frac{\text{tr}(\mathbf{\Lambda}^2)}{\text{tr}(\mathbf{\Lambda})^2}.$$

α	$\text{tr}(\mathbf{\Lambda})$	$\text{tr}(\mathbf{\Lambda}^2)$	ν
0.7	82.52	82.24	82.79
0.3	75.51	75.23	75.79
0.1	45.67	43.81	47.62

TABLE 9.1. One-moment and two-moment chi-square approximations. Comparing degrees of freedom for the ethanol dataset.

These moments can be used as the basis of a chi-square approximation. Letting $\nu = \text{tr}(\mathbf{\Lambda})^2/\text{tr}(\mathbf{\Lambda}^2)$,

$$\text{E}(\nu\frac{\hat{\sigma}^2}{\sigma^2}) = \nu$$

$$\text{var}(\nu\frac{\hat{\sigma}^2}{\sigma^2}) = 2\nu. \tag{9.5}$$

That is, the first two moments of $\nu\hat{\sigma}^2/\sigma^2$ match those of a chi-square distribution with ν degrees of freedom. This approximation was given for general quadratic forms by Satterthwaite (1946) and applied to local regression problems by Cleveland (1979) and section 6.2 of Katkovnik (1985). Simulations studying the accuracy are found in Cleveland and Devlin (1988).

The computation of $\text{tr}(\mathbf{\Lambda}^2)$ can be expensive. In light of (9.3) it is very tempting to use the simpler degrees of freedom approximation $\nu = \text{tr}(\mathbf{\Lambda})$ for the approximating chi-square distribution. Table 9.1 presents a small comparison of the degrees of freedom for the residual sum of squares for the ethanol dataset, using the one- and two-moment approximations. Three different smoothing parameters are used. Except at the smallest smoothing parameter, the three different degrees of freedom produce very similar numerical values, indicating that the one-moment approximation is adequate in this case.

9.1.1   Other Variance Estimates

Variance estimates other than the residual sum of squares can sometimes be useful. For example, if there is substantial replication of the x values, we can use the mean at each x value rather than a smooth function as the center when forming residuals. If all x values are duplicated, so that $x_{2i-1} = x_{2i}$ for all i, we could use

$$\hat{\sigma}^2 = \frac{1}{n}\sum_{i=1}^{n/2}(Y_{2i} - Y_{2i-1})^2.$$

The normalizing constant is derived by observing that $\text{E}(Y_{2i} - Y_{2i-1})^2 = 2\sigma^2$. Another class of variance estimates originated in the time series literature; see, for example, Section 3.4.4 of Anderson (1971). These estimates are based on difference sequences; for example,

$$\hat{\sigma}^2 = \frac{1}{6(n-2)} \sum_{i=2}^{n-1} (\nabla^2 y_i)^2 \tag{9.6}$$

where $\nabla^2 y_i = y_{i+1} - 2y_i + y_{i-1}$. In fact, this estimate can be interpreted as the normalized residual sum of squares from a three point moving average. But the difference interpretation is particularly intuitive and computationally simple.

The variance estimates considered so far are all quadratic forms in the data. This is convenient, since such estimates are relatively easy to compute and analyse. Moreover, the use of the residual sum of squares can be motivated as a likelihood estimate assuming a normal distribution. But quadratic forms are extremely sensitive to outliers, and for heavy tailed distributions can be quite inefficient.

In robust regression, scale estimates are frequently based on the absolute values of the residuals, rather than squared residuals. For example, the scale estimate based on the median absolute residual is

$$\hat{\sigma}_{\mathrm{MAD}} = 1.4826 \times \mathrm{median}_{1 \leq i \leq n} |Y_i - \hat{\mu}(x_i)|.$$

The leading factor, 1.4826, is chosen to ensure $\hat{\sigma}_{\mathrm{MAD}}$ is a consistent estimate of the residual standard deviation σ *when the Y_i are normally distributed.*

9.1.2 Nonhomogeneous Variance

One of the assumptions made when motivating a least squares procedure is that of a constant error variance. If this assumption is violated, the local least squares criterion (2.5) can be modified to

$$\sum_{i=1}^{n} \frac{w_i(x)}{\sigma_i^2} (Y_i - \langle a, A(x_i - x) \rangle)^2 \tag{9.7}$$

where $\sigma_i^2 = \mathrm{var}(Y_i)$. If the variances are known (at least up to a multiplicative constant), (9.7) yields the local parameter estimates

$$\hat{a} = (\mathbf{X}^T \mathbf{V} \mathbf{W} \mathbf{X})^{-1} \mathbf{X}^T \mathbf{W} \mathbf{V} Y \tag{9.8}$$

where $\mathbf{V}$ is a diagonal matrix with elements $1/\sigma_i^2$.

When the variances are unknown, some structure has to be assumed and the variances estimated. One common assumption is that the variance is a function of the mean,

$$\mathrm{var}(Y_i) = \sigma^2 V(\mu).$$

This model can be fitted using the quasi-likelihood procedure from Section 4.3.4; one interpretation of this procedure is an iterative algorithm in which one updates the variance estimate $V(\hat{\mu})$ at each step.

In other cases, the variance may not simply depend on the mean, but a more complicated function of the covariates x_i; $\sigma_i^2 = \sigma^2(x_i)$. In this case, the function $\sigma^2(x)$ must be estimated. An obvious approach to local variance estimation is to begin with raw estimates of the residual variance at each data point x_i, and smooth these against either $\hat{\mu}(x_i)$ (for quasi-likelihood) or x_i (for full local variance estimation). One specific implementation, used in Müller (1988) and Faraway and Sun (1995), is to use raw variance estimates based on differences; for example, $\hat{\sigma}_i^2 = (\nabla^2 y_i)^2/6$, and smooth these using a local average (rather than the global average used in (9.6)). Alternatively, one can use squared studentized residuals from a local fit. Ruppert, Wand, Holst and Hössjer (1997) use this approach in conjunction with local polynomial smoothers. For parametric regression models, see Carroll (1982). Exercise 9.2 gives an application to the motorcycle dataset.

An alternative local variance estimate, particularly suited for computation, is to use the deviations from the local polynomial in an estimate similar to (9.1). This gives

$$\hat{\sigma}^2(x) = \frac{\sum_{i=1}^{n} w_i(x)(Y_i - \langle \hat{a}, A(x_i - x)\rangle)^2}{\text{tr}(\mathbf{W}) - \text{tr}\left((\mathbf{X}^T\mathbf{W}\mathbf{X})^{-1}(\mathbf{X}^T\mathbf{W}^2\mathbf{X})\right)}. \tag{9.9}$$

The estimate (9.9) is easy to implement, since the weighted residual sum of squares is available as a by-product from the computation of $\hat{a}$. But it is based on a local constant approximation for $\sigma^2(x)$ within the smoothing window, which in turn requires $\hat{a}$ to be computed with a small bandwidth. This leads to a noisy estimate $\hat{\sigma}^2(x)$, which can be improved by smoothing.

Remark. One could iterate this procedure, alternately estimating the mean $\mu(x)$ and variance $\sigma^2(x)$. But there is usually little change in the estimates after the first iteration.

Example 9.1. Figure 9.1 displays measurements of the acceleration of a motorcycle that runs into a solid object. The data is from Schmidt, Mattern and Schüler (1981) and Härdle (1990). Clearly, the measurements exhibit nonhomogeneous variance.

To estimate the variance locally, first fit a local quadratic regression with a 10% nearest neighbor bandwidth, `alpha=0.1`:

```
> fit <- locfit(accel~time, data=mcyc, alpha=0.1)
```

From this fit, the numerator and denominator of (9.9) are found using the `predict()` function; note the use of `what="lik"` and `what="rdf"` for the two calls:

```
> x <- knots(fit, what="x")
> y <- -2 * predict(fit, what="lik", where="fitp")
> w <- predict(fit, what="rdf", where="fitp")
```

The local variance estimates are then smoothed using the gamma family; this is appropriate since σ^2 is a scale parameter:

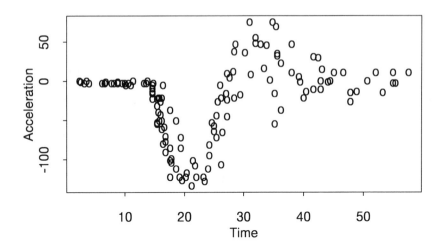

FIGURE 9.1. Motorcycle acceleration during a collision.

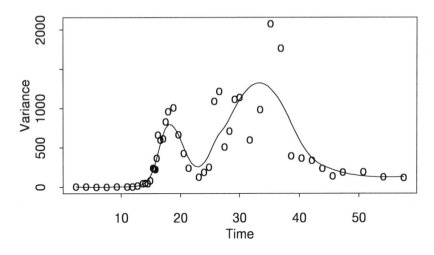

FIGURE 9.2. Local variance estimation for the motorcycle data.

```
> fitv <- locfit(y~x, weights=w, family="gamma", alpha=0.4)
> plot(fitv, get.data=T)
```

Figure 9.2 shows the local variance estimate. Note that successive values of $\hat{\sigma}^2(x)$, obtained from (9.9), are strongly correlated, so one should be careful of overinterpreting patterns in this plot.

9.1.3   Goodness of Fit Testing

One use of variance estimation is in goodness of fit testing. When a small bandwidth is used, the function estimate $\hat{\mu}(x)$, and hence the variance estimate $\hat{\sigma}^2$, will be nearly unbiased. As the bandwidth is increased, the bias increases, and this eventually translates into an increase in $\hat{\sigma}^2$.

Buckley and Eagleson (1989) used this as the basis for a graphical technique to decide whether the estimate shows lack of fit. More formally, the residual sum of squares can be used to form significance tests for lack of fit. These were studied in section 6.2 of Katkovnik (1985); Cleveland and Devlin (1988) and Bowman and Azzalini (1997). For ease of exposition, we focus on the hypotheses

$$\mathcal{H}_0 \;\; : \;\; \mu(x) = a + bx$$
$$\text{vs. } \mathcal{H}_1 \;\; : \;\; \text{otherwise,} \tag{9.10}$$

although the ideas are easily extended to other testing problems, such as comparing local fits with different bandwidths or testing significance of variables in multivariate local regression. Goodness of fit testing is closely related to model selection; criteria based on the residual sum of squares are used in both cases. But while model selection attempts to find models favored by the data, goodness of fit tests attempt to assess a statistical significance for features in a dataset.

Let $\hat{\alpha}_0 + \hat{\alpha}_1 x$ be a parametric least squares fit and $\hat{\mu}(x)$ be a local polynomial fit. Under these fits, compute the residual sums of squares:

$$\mathrm{RSS}_0 \;\; = \;\; \sum_{i=1}^{n}(Y_i - (\hat{\alpha}_0 + \hat{\alpha}_1 x_i))^2$$
$$= \;\; Y^T \Lambda_0 Y$$
$$\mathrm{RSS}_1 \;\; = \;\; \sum_{i=1}^{n}(Y_i - \hat{\mu}(x_i))^2$$
$$= \;\; Y^T \Lambda_1 Y$$

where Λ_0 and Λ_1 are defined by (9.2) for the global and local fits respectively. The F ratio for testing (9.10) is

$$F \;\; = \;\; \frac{(\mathrm{RSS}_0 - \mathrm{RSS}_1)/(\nu_0 - \nu_1)}{\mathrm{RSS}_1/\nu_1}$$
$$= \;\; \frac{(\nu_0 \hat{\sigma}_0^2 - \nu_1 \hat{\sigma}_1^2)/(\nu_0 - \nu_1)}{\hat{\sigma}_1^2}, \tag{9.11}$$

where $\nu_j = \mathrm{tr}(\Lambda_j)$. The distribution of the ratio (9.11) can be approximated using either the one-moment or two-moment chi-square approximations. An

α-level test of (9.10) rejects $\mathcal{H}_0$ if

$$F \geq F_{\nu_0 - \nu_1, \nu_1}(1 - \alpha).$$

The choice of bandwidth for the smooth fit $\hat{\mu}(x)$ remains. There have also been many other statistics proposed for the hypotheses (9.10); nearly all can be expressed as quadratic forms, and the F method can be used to approximate significance levels. References include Cox, Koh, Wahba and Yandell (1988), Azzalini, Bowman and Härdle (1989), Raz (1990), Azzalini and Bowman (1993), Hjellvik, Yao and Tjøstheim (1996) and Hart (1997).

One of the main considerations for choosing a test statistic is power: For a test of specified significance level, what is the probability that the null hypothesis is rejected when it is incorrect? In the local regression framework, there is no uniformly most powerful test, and one must compromise between power against the various models in the alternative hypothesis. Any legitimate test of (9.10) rejects $\mathcal{H}_0 : \mu(x) = a + bx$ if the observation vector $(Y_1, \ldots, Y_n)$ lies too far from the plane spanned by the vectors $(1, \ldots, 1)$ and $(x_1, \ldots, x_n)$. But departures from this plane lie in an $n-2$ dimensional space, and any test trades off power against alternatives in these $n-2$ dimensions. The F test will have most power against alternatives for which $E(\text{RSS}_0 - \text{RSS}_1)$ is large. Letting $\mu = (\mu(x_1), \ldots, \mu(x_n))^T$,

$$E(\text{RSS}_0 - \text{RSS}_1) = \sigma^2(\nu_0 - \nu_1) + \mu^T(\Lambda_0 - \Lambda_1)\mu. \tag{9.12}$$

Thus, the greatest power is in directions corresponding to the largest eigenvalues of $\Lambda_0 - \Lambda_1$. Typically, this matrix has a few eigenvalues close to 1 and many eigenvalues close to 0. The eigenvectors close to 1 correspond to smooth alternatives, against which the test has most power. If the bandwidth is small, then $\nu_0 - \nu_1$ is large and the power is spread out in many directions. Conversely, for a large bandwidth, the power of the test is concentrated in relatively few directions. This is most easily visualized by computing the decomposition in a few examples; see Exercise 9.4.

Although the F test was developed for regression models, it can be extended to likelihood models through the use of likelihood ratios or the difference of deviances.

Example 9.2. Consider the mine dataset used in Example 4.2. Table 9.2 shows the results of applying several tests to this dataset. The columns represent total residual deviance; residual degrees of freedom, change in deviance, and change in residual degrees of freedom. Finally, the P-value is computed using the chi-square distribution for $\Delta D(\hat{\theta})$ with $\Delta \nu$ degrees of freedom.

The first line fits the null model; a global constant mean. The second line fits a global log-linear model and compares with the global constant model. The resulting change in deviance is very significant. The third line fits the local linear model from Example 4.2. Again, the result is significant, so we conclude that the response (number of fractures) is nonlinear in this

	$D(\hat{\theta})$	ν	$\Delta D(\hat{\theta})$	$\Delta \nu$	P-value
constant	74.98	43.00			
extrp (linear)	48.62	42.00	26.36	1.00	2.8×10^{-7}
extrp (loc. lin.)	36.54	39.93	12.08	2.07	0.002
extrp*inb	32.42	34.24	4.11	5.69	0.62
extrp*seamh	35.34	34.09	1.18	5.83	0.97
extrp*time	30.69	35.42	5.84	4.50	0.26

TABLE 9.2. Mine dataset. Testing significance of terms for linear and local linear models in extrp, and bivariate models.

variable. The remaining three lines test three bivariate local linear models, against the univariate local linear model with the extrp predictor. These models show no significant improvement over the univariate model.

9.2   Interval Estimation

An interval estimate for the mean function $\mu(x)$ has the form $(L(x), U(x))$. The limits $L(x)$ and $U(x)$ are data-based quantities, chosen so that $L(x) \leq \mu(x) \leq U(x)$ with high confidence. The interval $(L(x), U(x))$ is a $(1 - \alpha)100\%$ **pointwise confidence interval** for $\mu(x)$ if

$$\sup_{\mu \in \mathcal{F}} P_\mu(L(x) \leq \mu(x) \leq U(x)) \geq 1 - \alpha. \tag{9.13}$$

Here, $\mathcal{F}$ denotes a suitable class of smooth functions.

Many questions depend on more than just single values of $\mu(x)$. For example, we may be interested in comparing mean responses at different levels of the covariate x or choosing a level of the covariates to maximize the mean response. Thus, we are also interested in constructing simultaneous confidence bands over a set $\mathcal{X}$, where $\mathcal{X}$ is typically taken to be a set bounding the predictors x_i. The band $\{(L(x), U(x)); x \in \mathcal{X}\}$ is a $(1 - \alpha)100\%$ **simultaneous confidence band** if

$$\sup_{\mu \in \mathcal{F}} P_\mu(L(x) \leq \mu(x) \leq U(x) \, \forall \, x \in \mathcal{X}) \geq 1 - \alpha.$$

9.2.1   Pointwise Confidence Intervals

To derive confidence intervals, the limits $(L(x), U(x))$ must be specified. Assuming ϵ_i are independent Gaussian random variables with mean 0 and variance σ^2, a local polynomial estimate $\hat{\mu}(x)$ has the distribution

$$\frac{\hat{\mu}(x) - \mathrm{E}(\mu(x))}{\sigma \|l(x)\|} \sim N(0, 1).$$

If the estimate is unbiased, so that $E(\hat{\mu}(x)) = \mu(x)$, confidence intervals may take the form

$$I_1(x) = (\hat{\mu}(x) - c\sigma\|l(x)\|, \hat{\mu}(x) + c\sigma\|l(x)\|), \qquad (9.14)$$

where c is chosen as the $(1 - \alpha/2)$ quantile of the standard normal distribution. When σ is replaced by the residual standard deviation $\hat{\sigma}$, c can also be chosen from the t distribution, with degrees of freedom defined by (9.5).

Formally, the interval $I_1(x)$ is a confidence interval for $E(\hat{\mu}(x))$. It is a confidence interval for $\mu(x)$ under the assumption $E\hat{\mu}(x) = \mu(x)$. In practice, this means undersmoothing, or choosing a small bandwidth for which one is willing to *assume* the bias is small, relative to the standard deviation of $\hat{\mu}(x)$.

An alternative is to adjust the intervals to allow for bias. If $b(x) = E\hat{\mu}(x) - \mu(x)$, a bias corrected confidence interval is

$$I_2(x) = (\hat{\mu}(x) - b(x) - c\hat{\sigma}\|l(x)\|, \hat{\mu}(x) - b(x) + c\hat{\sigma}\|l(x)\|).$$

Since $b(x)$ is unknown, a bias estimate $\hat{b}(x)$ is needed to form an estimated confidence intervals $\hat{I}_2(x)$. The most common approaches are based on the plug-in principle: either substitute $\hat{\mu}(x)$ directly into (2.33), or use derivative estimates in (2.34) or (2.41). But this doesn't solve the bias problem. Plug-in bias estimates simply amount to increasing the order of the fit. For example, in Exercise 2.5, the double smoothing bias correction converts a local constant estimate into a local quadratic. In this case an estimated $\hat{I}_2(x)$ is just a construction of an undersmoothed interval centered around the local quadratic estimate $\hat{\mu}(x) - \hat{b}(x)$. One has the additional problem that $\text{var}(\hat{\mu}(x) - \hat{b}(x))$ may be larger than $\text{var}(\hat{\mu}(x))$.

Another approach to bias adjustment is to focus on the class of smooth functions $\mathcal{F}$ in (9.13). For example, if $\mathcal{F}_\delta$ is defined to be the class of functions for which $|b(x)| \leq \delta$, then

$$I_3(x) = (\hat{\mu}(x) - \delta - c\hat{\sigma}\|l(x)\|, \hat{\mu}(x) + \delta + c\hat{\sigma}\|l(x)\|)$$

is a confidence interval for $\mu(x)$. Note the difference between $I_2(x)$ and $I_3(x)$: While $I_2(x)$ attempts to recenter the bands to allow for bias, $I_3(x)$ expands the bands. This type of expansion was used by Knafl, Sacks and Ylvisaker (1985) to construct simultaneous confidence bands. Sharper results, in which one attempts to adjust c rather than expanding by δ, were considered in Sun and Loader (1994). As with the bias corrected $I_2(x)$, adjusted intervals $I_3(x)$ require estimation of the bias. But the critical difference is that $I_3(x) \supset I_1(x)$, and thus the correction always improves the coverage probability.

9.2.2 Simultaneous Confidence Bands

The construction of simultaneous confidence bands is similar to confidence intervals: Begin with a band $\{I_1(x); x \in \mathcal{X}\}$ that is valid under the assumption of no bias, and then adjust the bands to allow for bias. The issues involved in bias estimation and adjustment are the same for simultaneous bands as they are for the pointwise intervals. Thus, our focus is on construction of bands under the no-bias assumption.

Under the assumption $\mu(x) = E(\hat{\mu}(x))$, the confidence band $\{I_1(x); x \in \mathcal{X}\}$ covers the true mean $\mu(x)$ if and only if

$$M_{\hat{\sigma}} = \sup_{x \in \mathcal{X}} \frac{|\hat{\mu}(x) - \mu(x)|}{\hat{\sigma}\|l(x)\|} \le c.$$

To find the critical value c, we need to find the distribution of $M_{\hat{\sigma}}$.

For linear regression, Scheffé (1959) showed that the distribution of M_{σ} is related to an F distribution when $\mathcal{X} = \mathcal{R}^d$. For more general problems - in particular, local regression - the exact distribution of $M_{\hat{\sigma}}$ is quite intractable and approximations must be used.

The first results approximating the distribution of $M_{\hat{\sigma}}$ for local fitting are those of Bickel and Rosenblatt (1973) who derive limiting extreme value distributions in the kernel density estimation setting. Extensions of this approach to kernel regression are considered in Eubank and Speckman (1993b) and elsewhere. This approach relies on the asymptotic stationarity of the process $(\hat{\mu}(x) - \mu(x))/\|l(x)\|$, so the usefulness is limited in finite sample size situations, where we have to respond to problems such as nonuniform designs and boundary effects.

For the one dimensional case, an alternative approach based on upcrossing methods was studied by Knafl, Sacks and Ylvisaker (1985). Their results were based on a discretized version of the following theorem.

Theorem 9.1 Suppose $\sigma = 1$ is known. Let $T(x) = l(x)/\|l(x)\|$ and assume $\{T(x); x \in \mathcal{X}\}$ is continuous. Then

$$P(M_1 \ge c) \le 2(1 - \Phi(c)) + \frac{\kappa_0}{\pi} e^{-c^2/2} \qquad (9.15)$$

where κ_0 is the length of the path $\{T(x); x \in \mathcal{X}\}$. If $\{T(x)\}$ is differentiable and $\mathcal{X} = [a, b]$,

$$\kappa_0 = \int_a^b \|T'(x)\| dx.$$

If the estimate $\hat{\sigma}^2$ has a χ^2 distribution with ν degrees of freedom, then

$$P(M_{\hat{\sigma}} > c) \le P(|t_\nu| > c) + \frac{\kappa_0}{\pi} \frac{1}{(1 + c^2/\nu)^{\nu/2}}.$$

The upper bound (9.15) is well known in the theory of Gaussian processes and can be derived in several different ways; see Aldous (1989) and references therein. The classical proof is based on Rice's formula (Rice 1939), which shows that the right-hand side of (9.15) is the expected number of crossings the process $\langle T(x), \epsilon \rangle$ makes of the boundaries at $\pm c$. Here, ϵ is assumed to have the multivariate $N(0, \mathbf{I})$ distribution. When c is large (corresponding to coverage probabilities close to 1), the number of upcrossings is usually either 0 or 1, and thus (9.15) provides an excellent probability approximation.

In multiple dimensions the simple upcrossing approach no longer works, and more sophisticated geometric arguments must be used. The main tools are formulae for the volumes of tubular neighborhoods of curves and surfaces derived by Hotelling (1939) and Weyl (1939). Modern work applying these methods to statistical problems includes Knowles and Siegmund (1989), Naiman (1990) and Sun (1993). The application to local regression was first studied in Sun and Loader (1994). These results yield the approximation

$$P(M_{\hat{\sigma}} \geq c) \approx \kappa_0 \frac{\Gamma(\frac{d+1}{2})}{\pi^{(d+1)/2}} P(F_{d+1,\nu} > \frac{c^2}{d+1}) + \frac{\zeta_0}{2} \frac{\Gamma(\frac{d}{2})}{\pi^{d/2}} P(F_{d,\nu} > \frac{c^2}{d})$$

$$+ \frac{\kappa_2 + \zeta_1 + m_0}{2\pi} \frac{\Gamma(\frac{d-1}{2})}{\pi^{(d-1)/2}} P(F_{d-1,\nu} > \frac{c^2}{d-1}) \qquad (9.16)$$

where κ_0, ζ_0, κ_2, ζ_1 and m_0 are certain geometric constants. In particular, κ_0 represents the area, or volume, of the set $\mathcal{I} = \{T(x) : x \in \mathcal{X}\}$. Explicitly,

$$\kappa_0 = \int_{\mathcal{X}} \det{}^{1/2}[\langle T_i(x), T_j(x)\rangle] dx,$$

where $T_i(x)$ denotes the partial derivative of $T(x)$ in the ith direction and $[\cdot]$ denotes a matrix with the given (i, j)th elements. ζ_0 is the volume of the boundary of $\mathcal{X}$ and m_0 is related to the corners of $\mathcal{X}$. κ_2 and ζ_1 are more complicated constants related to the curvature of $\mathcal{X}$ and the boundary of $\mathcal{X}$ respectively. When $d = 2$ it is known that $\kappa_0 + \kappa_2 + \zeta_1 + m_0 = 2\pi$ for simple sets $\mathcal{X}$, such as rectangles.

Computation of κ_0 requires numerical integration and is slow in multiple dimensions or at small bandwidths. But clearly κ_0 must be strongly related to the degrees of freedom of the fit. For example, for one dimensional $\mathcal{X}$, if $T(x)$ rotates through 90 degrees as x changes, this amounts to adding one degree of freedom to the fit and adding $\pi/2$ to κ_0. Thus, to a very rough approximation,

$$\kappa_0 \approx \frac{\pi}{2}(\nu_1 - 1)$$

where ν_1 is the trace of the hat matrix.

Improved approximations are obtained by considering the case of a uniform design. When the x_i are equally spaced, one can obtain a limiting relation similar to (2.43). Table 9.3 displays some approximate results for the tricube weight function.

The dominant term can be easily deduced. Considering the simplification in Exercise 9.3 and ignoring boundary effects, symmetry implies that

d	Degree	Limit
1	0	$-1.400 + 2.0602\nu_1$
1	1	$-2.181 + 2.0602\nu_1$
1	2	$-1.764 + 1.8835\nu_1$

TABLE 9.3. Approximate values of κ_0 as a function of the fitted degrees of freedom ν_1 for different local polynomial degrees.

$\langle l(x), l_i(x) \rangle = 0$ for all i, and $\langle l_i(x), l_j(x) \rangle = 0$ for $i \neq j$. Thus

$$\kappa_0 \approx \sqrt{\frac{\operatorname{var}(\hat{\mu}'(x))}{\operatorname{var}(\hat{\mu}(x))}}.$$

Using (2.39) and a similar expression for the local slope then yields

$$\kappa_0 \approx \frac{1}{h} \sqrt{\frac{\int W'(v)^2 dv}{\int W(v)^2 dv}}.$$

Recalling the degrees of freedom approximation (2.42), we obtain

$$\kappa_0 \approx \sqrt{\frac{\int W'(v)^2 dv}{W(0)}} \kappa_0.$$

9.2.3 Likelihood Models

The preceding development of confidence bands has been for the local regression model, under the normal assumption. For likelihood models, confidence intervals should ideally take into account the underlying family of distributions But the theory for deriving such intervals seems quite intractable, and we must rely on methods based on normal assumption, using the approximate variance from Section 4.4.3.

A problem that occurs in likelihood models is that $\operatorname{var}(\hat{\mu}(x)$ usually depends on the unknown parameter $\mu(x)$, and simply substituting an estimate (which (4.19) does, through the estimation of $\mathbf{V}$) may not be satisfactory, particularly in boundary cases. A standard example is the estimation of a binomial parameter: $Y \sim \operatorname{Bin}(n, p)$; $\hat{p} = Y/n$ and the 95% confidence interval using a normal approximation is $\hat{p} \pm 1.96\sqrt{\hat{p}(1-\hat{p})/n}$. This obviously fails to produce sensible results, if we happen to observe $Y = 0$. Using the logistic link function $\theta = \log(p/(1-p))$ doesn't help: on this scale, $\hat{\theta} = -\infty$, and the variance estimate is also infinite.

The simple solution, within the framework of normal approximations, is to use the variance stabilizing link. Under this link, the variance of $\hat{\theta}(x)$ is, at least asymptotically, independent of the true parameter $\theta(x)$, leading to confidence intervals whose widths depend only on the design points x_i.

Example 9.3. Consider local logistic regression with the binomial family; $P(Y_i = 1) = p(x_i)$, and the arcsin link, $\theta(x) = \sin^{-1}(\sqrt{p(x)})$. Inverting this relation yields $p(x) = \sin^2(\theta(x))$ and $1 - p(x) = \cos^2(\theta(x))$. Thus, the log-likelihood is

$$l(Y_i, \theta_i) = 2Y_i \log \sin(\theta_i) + 2(1 - Y_i) \log \cos(\theta_i)$$

and with some manipulation, one finds

$$\ddot{l}(Y_i, \theta_i) = -2 \left(\frac{Y_i}{\sin^2(\theta_i)} + \frac{1 - Y_i}{\cos^2(\theta_i)} \right) = -2 \left(\frac{Y_i}{p_i} + \frac{1 - Y_i}{1 - p_i} \right).$$

Since $E(Y_i) = p_i$, $E(\ddot{l}(Y_i, \theta_i)) = -4$, independent of θ_i. Thus, the arcsin link is variance stabilizing for the binomial family. The variance approximation (4.19) reduces to

$$\mathrm{var}(\hat{\theta}(x)) \approx \frac{1}{4} \|l(x)\|^2,$$

where $l(x)$ is the local regression weight diagram, defined by (2.12).

Variance stabilizing links for most of the families supported in LOCFIT were indicated in Table 4.1.

9.2.4 Maximal Deviation Tests

In section 9.1.3, goodness of fit tests using quadratic forms and approximate F tests were introduced. These tests may not be informative, since rejecting the null model provides no information as to what form the lack of fit takes. Maximal deviation tests attempt to address this problem by determining whether individual features are significant.

Suppose we are interested in testing the null hypothesis that $\mu(x)$ is constant. We estimate $\mu(x)$ using a local polynomial (linear or higher order) fit. Under the null hypothesis, the local slope estimate $\hat{\mu}'(x)$ has mean 0. As a test statistic, we can consider the scaled maximum of $\hat{\mu}'(x)$:

$$M_{\hat{\sigma}} = \sup_{x \in \mathcal{X}} \frac{|\hat{\mu}'(x)|}{\hat{\sigma}\sqrt{\mathrm{var}(\hat{\mu}'(x))}} \tag{9.17}$$

for a suitable set $\mathcal{X}$. The problem of computing critical values for this test statistic is closely related to the simultaneous confidence band problem, and Theorem 9.1 can be used to compute approximate critical values. Note that by formulating the problem as a hypothesis test, the bias problem is avoided. Under the null hypothesis of a constant mean, the local slope $\hat{\mu}'(x)$ has mean 0, regardless of the bandwidth.

Example 9.4. We return to the Weibull fit of the heart transplant dataset from Example 7.8. The question is whether the decrease in mean

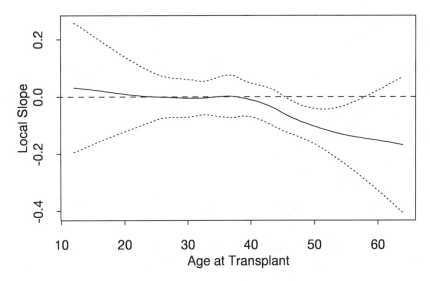

FIGURE 9.3. Testing for a significant feature: Local slopes and confidence bands for the heart transplant data.

survival times on the right is significant. We perform this test by looking at the local slope from the fit, and consider the test statistic (9.17).

First, we need to compute the fit for the local slope by setting `deriv=1` in the `locfit()` call. Then critical values for the simultaneous bands are computed by the `kappa0()` function and substituted on the fit. Finally, we plot the fit and global confidence bands:

```
> fit <- locfit(I((surv+0.5)^0.625)~age, cens=cens,
+   data=heart, alpha=0.8, family="gamma", deriv=1)
> crit(fit) <- kappa0(fit)
> plot(fit, band="global", xlab="Age at Transplant",
+   ylab = "Local Slope")
> abline(h=0)
```

Figure 9.3 shows the local slope estimate and confidence bands. For $45 <$ `age` < 55, the confidence bands do not include 0, indicating the decrease is significant in this region. While the slope continues to decrease as `age` increases to 65, the confidence bands also get wider due to boundary variability.

The statistic (9.17) provides a legitimate test for any bandwidth h, and different bandwidths will provide tests with different power. Thus, different information can be obtained by looking at test statistics computed by varying the bandwidth. Of course, this multiple testing doesn't preserve the significance level of a single test. But the bandwidth h can be treated as another dimension of the random field, leading to the test statistic

$$M'_{\hat{\sigma}} = \sup_{x \in \mathcal{X}, h_0 \leq h \leq h_1} \frac{|\hat{\mu}'(x, h)|}{\hat{\sigma} \sqrt{\mathrm{var}(\hat{\mu}'(x, h))}}$$

for a suitable range $h_0 \leq h \leq h_1$ of bandwidths. Critical values for $M'_{\hat{\sigma}}$ can be computed using (9.16), except that the integral defining κ_0 must now be taken over both x and h. This approach was studied for a continuously observed process by Siegmund and Worsley (1995).

9.3   Exercises

9.1 Let Λ be a symmetric non-negative definite $n \times n$ matrix.

   a) Show $\mathrm{tr}(\Lambda)^2/\mathrm{tr}(\Lambda^2) \leq n$, with equality if and only if Λ is a multiple of the identity matrix.

   b) Show $\mathrm{tr}(\Lambda)^2/\mathrm{tr}(\Lambda^2) \geq 1$, with equality if and only if Λ has rank 1.

   Hint: First show for diagonal Λ, then use the eigenvalue decomposition.

9.2 An alternative to smoothing the local variance estimates in Example 9.1 is to smooth the studentized squared residuals.

   a) Compute the fit for the motorcycle dataset with `alpha=0.1` from Example 9.1. Compute and plot the squared residuals against the `time` variable.

   b) Compute a local variance estimate using local quadratic fits with `alpha=0.4`. Plot the fit and compare with Figure 9.2. Try using both `family="gamma"` and `family="gauss"`. Which family appears to produce the better variance estimate?

9.3 Let

$$\Lambda = \begin{pmatrix} a & v^T \\ v & \mathbf{M} \end{pmatrix}$$

where $a > 0$, v is a vector and $\mathbf{M}$ is a symmetric matrix. Show

$$\det \Lambda = a \det(\mathbf{M} - vv^T/a).$$

Use this result to show

$$\det[\langle T_i(x), T_j(x) \rangle]$$
$$= \frac{1}{\|l(x)\|^{2d+2}} \det \begin{pmatrix} \|l(x)\|^2 & [\langle l(x), l_i(x) \rangle] \\ [\langle l_i(x), l_i(x) \rangle] & [\langle l_i(x), l_j(x) \rangle] \end{pmatrix}.$$

where $T_i(x)$ are as defined in Section 9.2.2. Let $\mathbf{R} = [r_{i,j}]$ be the right triangular matrix from the QR-decomposition of the matrix $(\, l(x) \quad l_1(x) \quad \ldots \quad l_d(x)\,)$. Show

$$\det[\langle T_i(x), T_j(x) \rangle]^{1/2} = \prod_{j=2}^{d+1} \frac{r_{i,i}}{r_{1,1}}.$$

9.4 This exercise investigates the power of likelihood ratio type tests, based on eigen-decompositions of the quadratic forms.

a) For the ethanol dataset, compute the hat matrix for linear regression:

```
> L0 <- t(hatmatrix(NOx~E, data=ethanol, ev="data",
+    kern="parm",deg=1))
```

Also compute the hat matrix L1 for a local quadratic fit with alpha=0.3.

b) Compute the matrices $\mathbf{\Lambda}_0$ and $\mathbf{\Lambda}_1$ appearing in (9.12) and the eigenvalues and eigenvectors of the difference $\mathbf{\Lambda}_0 - \mathbf{\Lambda}_1$.

c) Plot the eigenvalues. For the largest eigenvalues (those close to 1) plot the corresponding eigenvectors, using ethanol$E for the x-axis. Also plot some of the eigenvectors corresponding to eigenvalues close to 0.

d) Repeat this exercise for a range of other smoothing parameters α.

e) Repeat this exercise for other goodness of fit tests based on quadratic forms, such as Härdle and Mammen (1993) or Hjellvik, Yao and Tjøstheim (1996) (this may involve a lot of programming, to get the correct quadratic form matrices for these tests). Which tests seem most suited to smooth alternatives?

10
Bandwidth Selection

In earlier chapters, statistics such as cross validation, CP and AIC have been introduced as tools to help assess the performance of local polynomial fits. One goal is automatic bandwidth and model selection: an algorithm that takes the data as input and produces the best local polynomial fit as output.

Unfortunately this goal is unattainable, since there is often considerable uncertainty in data and it is unclear what the best fit should be. Figure 2.3 showed four local quadratic fits to the ethanol dataset, each computed with a different smoothing parameter. As discussed in Section 2.2.1, the largest smoothing parameter, $\alpha = 0.8$, clearly oversmooths. Choosing among the remaining fits is indecisive: While changing the smoothing parameter from $\alpha = 0.6$ to $\alpha = 0.2$ results in fits showing different structure in the data, making a definitive statement as to which fit is best is quite impossible. That is, there is wide uncertainty in the data.

Similar behavior was seen for the Old Faithful geyser density in Figures 5.7 and 5.8. The largest smoothing parameter, $\alpha = (0.1, 1.2)$, appears to oversmooth the data, trimming the left peak and filling in the valley. The smaller smoothing parameters showed increasing amounts of structure, but no definitive statement can be made as to which fit is best.

What should be expected from a bandwidth selector on these datasets? In both cases, the largest bandwidth should be rejected, since the resulting estimates do not fit the data. At the smaller bandwidths, there is uncertainty in the problem, and a bandwidth selector should be indecisive: It is unclear which fit is best, and a bandwidth selector should reflect this uncertainty and not attempt to make a definitive judgment.

This is precisely the behavior indicated in the GCV plot in Figure 2.7 and AIC plot in Figure 5.8 for these two examples. In both cases, the largest bandwidths (smallest degrees of freedom) result in high GCV and AIC scores. For the ethanol dataset, the GCV scores are $0.184, 0.123, 0.109$ and 0.111 for $\alpha = 0.8, 0.6, 0.4$ and 0.2 respectively, which is almost flat for the smaller smoothing parameters. Likewise, the AIC plot in Figure 5.8 is almost flat for the smaller smoothing parameters. Precisely, this *reflects the uncertainty in the data*: Inadequate fits are rejected, while those we aren't sure about produce similar results.

Despite the well motivated and intuitive behavior of the cross validation and AIC methods, these procedures have received considerable criticism in recent literature. For example, Marron (1996) states that cross validation methods are "widely accepted as 'too noisy' ". Jones, Marron and Sheather (1996) categorize cross validation as 'first generation methods', and claim other plug-in, or 'second generation methods' have "a quantum leap in terms of performance (both theoretical and practical)".

In this chapter the bandwidth selection issue is explored in depth. In particular, we carefully distinguish between two factors that contribute to the performance of bandwidth selectors:

1. What are the apriori assumptions being made by the bandwidth selector?

2. What is the ability of the bandwidth selector to resolve uncertainty in the data?

When these factors are properly identified, we find that cross validation methods far outperform the 'second generation' methods. Simply, the noise and variability in cross validation reflect uncertainty in the data. Other methods reflect this uncertainty in different ways; in particular, by missing important aspects of difficult smoothing problems and selecting estimates which clearly do not fit the data. The 'second generation' methods generally have less ability to resolve uncertainty in the data.

The distinction between the questions stated above is important, but requires careful thought. Some of the points made in this chapter are quite subtle, and proper understanding will require careful reading. Some of the examples are from Loader (1999).

10.1   Approaches to Bandwidth Selection

10.1.1   *Classical Approaches*

Classical approaches to bandwidth selection are extensions of model selection methods used in parametric statistics. Diagnostic tools introduced

throughout this book, such as cross validation methods, AIC, CP and goodness of fit tests, fall into this category.

Several variants of these ideas exist; in particular, many different penalty
functions have been considered. Kooperberg and Stone (1992) and Stone,
Hansen, Kooperberg and Truong (1997), in the context of spline models,
recommend modifying the AIC criterion (4.11) by increasing the penalty
$2\nu_1$. The choice $\log(n)\nu_1$ corresponds to the BIC criterion, introduced by
Schwarz (1978). The increased penalty ensures that larger smoothing parameters are selected, and smoother fits are produced.

Other authors have proposed methods that are asymptotically equivalent
to AIC or cross validation but have different finite sample penalties. One
example is Rice's T statistic (Rice 1984):

$$\mathrm{RT}(\hat{\mu}) = \frac{\sum_{i=1}^{n}(Y_i - \hat{\mu}(x_i))^2}{n - 2\nu_1}.$$

This is very similar to the generalized cross validation statistic (2.21).
Providing the fitted degrees of freedom ν_1 is small relative to the sample
size n, $(n - \nu_1)^2/n \approx n - 2\nu_1$, so GCV and RT are almost identical numerically. The difference occurs at large degrees of freedom. Since $\mathrm{RT}(\hat{\mu})$
has a singularity at $\nu_1 = n/2$, the statistic has *apriori rejected* any fit with
$n/2$ degrees of freedom and strongly penalized against any fits with close
to $n/2$ degrees of freedom.

It must be remembered that changing the penalty is purely an expression of prior assumption and in no way helps to resolve uncertainty in the
data. For example, the $\mathrm{RT}(\hat{\mu})$ and $\mathrm{GCV}(\hat{\mu})$ statistics have *exactly* the same
random component, namely the residual sum of squares. Therefore the two
statistics have *exactly* the same ability to resolve uncertainty in the data,
even though $\mathrm{RT}(\hat{\mu})$ will generally prefer larger bandwidths.

10.1.2 Plug-In Approaches

The fundamental feature of plug-in selectors is direct estimation of the
bias of local estimates. The idea first appeared in Woodroofe (1970) and
was developed by Scott, Tapia and Thompson (1977) and Tsybakov (1987),
among others. The approach has been widely promoted in recent literature:
for example Park and Marron (1990), Sheather and Jones (1991), Gasser,
Kneip and Köhler (1991) and Ruppert, Sheather and Wand (1995). Jones,
Marron and Sheather (1996) describe plug-in algorithms as "second generation methods", and suggest the Sheather-Jones algorithm should be "a
benchmark for good performance".

Plug-in selectors are most widely developed for kernel density estimation
with a constant bandwidth h. The bias of a density estimate $\hat{f}_h$ is written
as a function of the unknown f, and usually approximated through a Taylor
series expansion. An estimate of f is then plugged in to derive estimates of

the bias and hence goodness of fit. One then chooses a bandwidth h that minimizes this estimated goodness of fit.

The most commonly used goodness of fit measure is mean integrated squared error:

$$\text{MISE}(\hat{f}_h, f) = \text{E} \int_{-\infty}^{\infty} (\hat{f}_h(x) - f(x))^2 dx$$

$$= \int_{-\infty}^{\infty} \text{bias}(\hat{f}_h(x))^2 dx + \int_{-\infty}^{\infty} \text{var}(\hat{f}_h(x)) dx. \quad (10.1)$$

Using the approximations of Exercise 5.1, an asymtotic approximation is

$$\text{MISE}(\hat{f}_h, f) = \frac{h^4}{4} \left(\int v^2 W(v) dv \right)^2 \int_{-\infty}^{\infty} f''(x)^2 dx$$

$$+ \frac{1}{nh} \int W(v)^2 dv + o(h^4 + (nh)^{-1}) \quad (10.2)$$

where for simplicity we assume $\int W(v) dv = 1$. This result is theorem 6.1 of Scott (1992). Minimizing this approximation over h yields an approximation to the best bandwidth, stated in the following theorem.

Theorem 10.1 Let h_{opt} be the bandwidth minimizing $\text{MISE}(\hat{f}_h(x), f(x))$. Then

$$h_{\text{opt}}^5 = \frac{\int W(v)^2 dv}{n(\int v^2 W(v) dv)^2 \int f''(x)^2 dx}(1 + o(1)) \quad (10.3)$$

as $n \to \infty$.

The bandwidth provided by Theorem 10.1 depends on the unknown quantity $\int f''(x)^2 dx$, that must be estimated. Choose a pilot bandwidth k, and construct a 'pilot' estimate of the second derivative:

$$\hat{f}_k''(x) = \frac{1}{nk^3} \sum_{i=1}^{n} W'' \left(\frac{X_i - x}{k} \right). \quad (10.4)$$

Then, substituting $\hat{f}_k''(x)$ into (10.3) provides an estimate of h_{opt}.

Example 10.1. Figure 10.1 displays three estimates of $f''(x)^2 dx$ for the Old Faithful dataset. In each case, a local quadratic fit with the Gaussian kernel and identity link is used. The corresponding estimates of $\int f''(x)^2 dx$ and bandwidth h chosen by (10.3) are shown in Table 10.1. The important point here is the sensitivity of plug-in selection to the choice of pilot bandwidth: Doubling the pilot bandwidth k changes the estimate of $\int f''(x)^2 dx$ by a factor of 10, and the selected h changes by a factor of 1.6.

The solid line in Figure 10.2 displays the full relationship between the pilot bandwidth k and selected bandwidth h for the Old Faithful dataset.

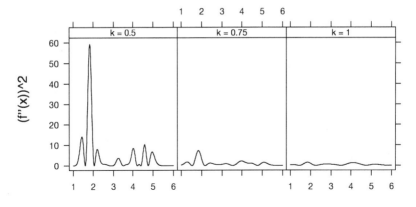

Eruption Duration

FIGURE 10.1. Estimating $f''(x)^2 dx$ for the Old Faithful dataset, using three pilot bandwidths k.

k	$\int \hat{f}_k''(x)^2 dx$	h
0.50	23.8	0.40
0.75	5.5	0.54
1.00	2.3	0.65

TABLE 10.1. Pilot bandwidths, estimates of $\int f''(x)^2 dx$ and selected bandwidths for the Old Faithful dataset.

Clearly plug-in bandwidth selection alone does not *solve* the bandwidth problem, but *replaces* the problem with the problem of choosing pilot bandwidths. In light of the discussion in Section 6.1.1, it is clear that the second derivative relies on a good local quadratic approximation. If k is so large as to smooth out features, the second derivative will be underestimated, and h_{opt} overestimated.

Several solutions have been proposed for the pilot bandwidth problem. The most common ideas center around assumed relations between h and k. Some specific ideas include:

- Gasser, Kneip and Köhler (1991), working in the regression setting, assume the relation

$$k(h) = hn^{1/10}.$$

This relation is shown by the long dashed curve in Figure 10.2. The selected bandwidth, $h = 0.670$, is defined by the intersection of the assumed and plug-in relations.

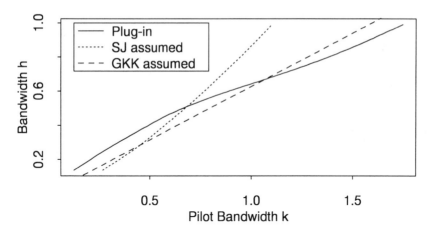

FIGURE 10.2. Plug-in relation (solid) between pilot bandwidth and selected bandwidth, and two assumed relations for the Old Faithful dataset.

- The Sheather-Jones method (SJPI). The assumed relation has the form

$$k(h) = Ch^{5/7},$$

  where C depends on the data, but not the bandwidth h. See Sheather and Jones (1991) for a more complete description. This relation is shown by the short dashed curve in Figure 10.2. Again, the selected bandwidth, $h = 0.516$, is defined by the intersection of the assumed and plug-in relations.

- Biased cross validation (Scott and Terrell 1987). The name is a misnomer, since the algorithm has little to do with cross validation. Instead of targeting the bandwidth (10.3), Scott and Terrell work directly with the asymptotic MISE (10.2). For each h, the BCV$(\hat{f}_h)$ criterion is obtained by substituting

$$\hat{B}(\hat{f}_h, W) = \int_{-\infty}^{\infty} \hat{f}_h''(x)^2 dx - \frac{1}{nh^5} \int_{-\infty}^{\infty} W''(v)^2 dv$$

  for $\int f''(x)^2 dx$. The bandwidth is then selected as the minimum (or a local minimum) of BCV$(\hat{f}_h)$.

10.2 Application of the Bandwidth Selectors

In this section we study the performance of the bandwidth selectors in the kernel density estimation setting. Both real data examples and simulations are used.

10.2.1 Old Faithful

The Old Faithful geyser dataset is the most widely used example in the bandwidth selection literature. Following authors such as Sheather and Jones (1991) and Scott (1992) we apply the kernel density estimate (5.6) with the Gaussian kernel. Applying the likelihood cross validation criterion (5.12) gives $h = 0.315$. The AIC criterion (5.15) gives $h = 0.162$. The LSCV criterion (5.17) yields $h = 0.249$.

We have already seen that the SJPI method yields $h = 0.516$ and the GKK method $h = 0.670$. Biased cross validation gives $h = 0.705$. Other plug-in selectors produce similar results: Chiu (1991) selects $h = 0.537$. Sheather (1992) reports $h = 0.570$ using a plug-in method of Park and Marron (1990) and the rather larger $h = 1.235$ using the method of Hall, Sheather, Jones and Marron (1991). Mostly, the plug-in selectors agree with Silverman (1986), who smoothed the data visually and selected $h = 0.625$.[1]

Selector	h	$h/2.5$	d.f.
AIC	0.162	0.065	13.1
LSCV	0.249	0.099	9.0
LCV	0.315	0.126	7.2
SJPI	0.516	0.206	4.4
Chiu	0.537	0.215	4.2
Park-Marron	0.570	0.228	3.9
Silverman	0.625	0.250	3.6
GKK	0.670	0.268	3.4
BCV	0.705	0.282	3.3
Hall et al.	1.235	0.494	2.1

TABLE 10.2. Bandwidths selected for the Old Faithful dataset: Three classical selectors, six plug-in selectors, and the visual selection of Silverman. The h column uses the LOCFIT definition of the Gaussian kernel; the $h/2.5$ column uses the definition in the original sources.

Table 10.2 summarizes the selected bandwidths. There is a large discrepancy between the approaches; the classical selectors select fits with smaller bandwidths and larger degrees of freedom.

Figure 10.3 shows the density estimates produced by some of these selectors. Clearly the fits are very different, corresponding to the very different bandwidths selected. Given the substantial discrepancy between the bandwidth selectors in the previous section, which is right? Existing literature has strongly rejected the LSCV fit. For example, Scott (1992, page

---

[1]**Important:** The bandwidths reported here are 2.5 times those reported in the original sources, reflecting the factor 2.5 in the definition of the Gaussian kernel given in Table 3.1 and used throughout this book.

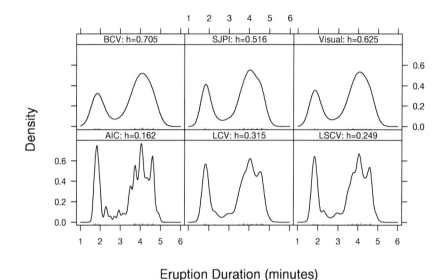

Density

Eruption Duration (minutes)

FIGURE 10.3. Kernel density estimates for the Old Faithful dataset.

172), says the LSCV fit is "clearly undersmoothed given the sample size". Sheather (1992, page 234), also says "it appears (LSCV) undersmooths the data" but goes on to add "the lack of agreement between these estimates is a little alarming". On the other hand, Figure 5.8 shows both the AIC and LSCV criteria fairly strongly rejecting fits with fewer than five degrees of freedom.

With a real dataset we can never be sure, but substantial insight can be gained by simulating from distributions that look similar the candidate distributions and seeing how the bandwidth selectors perform on these simulated datasets. Random variables are generated from the density

$$f_\sigma(x) = \frac{1}{107\sigma} \sum_{i=1}^{107} \phi\left(\frac{x - X_i}{\sigma}\right)$$

where $X_1, \ldots, X_{107}$ are the Old Faithful observations. This distribution can be generated by resampling (with replacement) from the Old Faithful observations and adding independent normal observations with standard deviation σ. Samples of size 107 are drawn, and two values of σ are considered: $\sigma_0 = 0.208$ and $\sigma_1 = 0.069$.

For normal mixtures, the mean integrated squared error can be obtained in a closed form; see Alekseev (1982), Taylor (1989) and Marron and Wand (1992). The MISE-minimizing bandwidth can then be found through a numerical computation. This leads to the bandwidths $h_0 = 0.516$ for $\sigma = \sigma_0$ and $h_1 = 0.249$ for $\sigma = \sigma_1$.

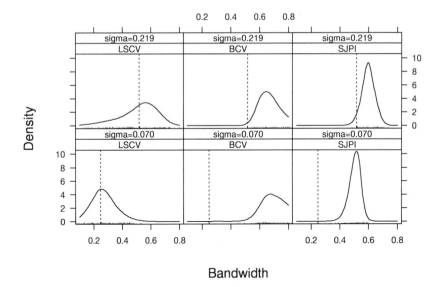

Bandwidth

FIGURE 10.4. Bandwidth selection for the resampled Old Faithful simulations. Estimated densities of the bandwidths selected by the LSCV, BCV and SJPI selectors are shown. The target bandwidth in each case is shown by the vertical line.

LOCFIT provides a function `kdeb()` that implements some of the bandwidth selectors. The simulations are implemented by calling this function in a loop:

```
> nsim <- 1000
> sigma <- 0.208
> Z <- matrix(nrow=nsim,ncol=3)
> meth <- c("LSCV","BCV","SJPI")
> for (i in 1:nsim)
+ { x <- sample(geyser,107,replace=T)+rnorm(107,sd=sigma)
+   Z[i,] <- kdeb(x,0.01,1.0,meth=meth)
+ }
```

For each value of σ, 1000 simulations are performed. In each of the six cases (three bandwidth selectors times two standard deviations), the density of the 1000 selected bandwidths is estimated using a local log-quadratic density estimate.

The results in Figure 10.4 are extremely interesting and informative. The LSCV selector is indeed the most variable, particularly for $\sigma = 0.219$. But in both cases it is correctly centered, with the peak of the estimated density being close to the target value. The BCV selector is less variable, but nearly always selects bandwidths that are much too large. It shows almost no ability to distinguish between the two models; the densities of the selected bandwidths are similar for both values of σ. The SJPI selector is the least variable of the selectors, but shows only modest ability to distinguish between the two values of σ. It also substantially oversmooths, particularly for $\sigma = 0.07$.

Returning to the questions posed in the introduction to this chapter, we can now say that the selection of larger bandwidths by the plug-in methods has nothing to do with resolution of uncertainty in the data. Rather, it is entirely due to apriori assumptions implicit in the selectors: the larger bandwidths are selected regardless of the truth.

These simulations cast considerable doubt on the performance of the plug-in selectors on the original dataset in Figure 10.3. What should be apparent, both from the residual analysis in Figure 5.7 and the second derivative estimates in Figure 10.1, is that the sharpness of the left peak is critical to the bandwidth selectors. Some probability calculations at the left boundary (Exercise 10.1) provide further support for the smaller bandwidths produced by cross validation methods.

There is one more important conclusion to be drawn from this section. *Looking at fits alone provides no basis for rejecting fits as undersmoothed.* Selecting the best fit is a trade-off between bias and variance. Plotting the fit, by itself, shows only variance in the estimate. Bias requires careful comparison of the fit with the data. The graphical tools used in Section 5.3 - residual plots and AIC plots - help provide a visual display of the balance between bias and variance.

10.2.2   The Claw Density

The claw density is defined as

$$f(x) = 0.5\phi(x) + \sum_{i=-2}^{2} \phi(10(x - i/2)).$$

This consists of an underlying standard normal component with five claws superimposed. This density is from Marron and Wand (1992).

Example 10.2. Figure 10.5 shows a sample of size 54 from the claw density, along with two kernel density estimates. The first fit is produced by

```
> x <- seq(-3.5, 2.7, length=200)
> y <- dnorm(x,-1,0.1) + dnorm(x,-0.5,0.1) + dnorm(x,0,0.1)
+    + dnorm(x,0.5,0.1) + dnorm(x,1,0.1)
> y <- (y+5*dnorm(x))/10
> fit1 <- locfit(~claw54, deg=0, kern="gauss", ev="grid",
+    mg=100, alpha=c(0,0.315), flim=c(-3.5,2.7))
> plot(fit1, get.data=T, main="h=0.315", ylim=c(0,max(y)))
> lines(x, y, lty=2)
```

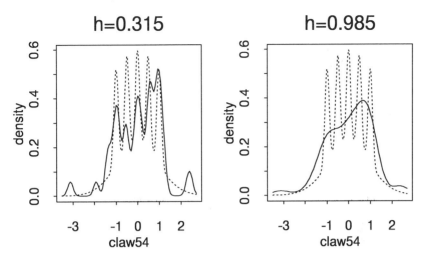

FIGURE 10.5. Density estimates for a sample from the claw density.

and similar code for $h = 0.985$.

The smaller bandwidth, $h = 0.315$, shows the five claws, albeit with a lot of noise. This is to be expected, given the small sample size. The larger bandwidth, $h = 0.985$, produces a smooth density estimate, which misses the claws completely.

This problem is particularly challenging for bandwidth selectors, since the interesting structure (the five claws) is difficult to see. Without any selectors, one could easily conclude that the estimate on the right ($h = 0.985$) is reasonable. *Thus, the performance of bandwidth selectors, in flagging features that might otherwise be missed, is particularly crucial in this type of example.*

Since the bandwidth selectors we consider in this chapter target the mean integrated squared error, we must consider how this measure behaves as a function of the bandwidth h and sample size n. When n is small, the estimate $\hat{f}_h(x)$ is noisy for small h, and the claws are impossible to detect. The best h will correspond to detecting the global structure.

As n increases, the noise is reduced and the claws are more detectable. The sample size $n = 54$ turns out to be critical. When n is close to 54, the MISE curve has two local minima: A large h corresponding to the global structure, and a small h corresponding to the claws. The cross-over value is $n = 54$, where the two local minima are the same height. Thus, at this sample size, we should expect a bandwidth selector to have about a 50% chance of finding the claws.

As n increases further, the claws become increasingly dominant. A second critical sample size is $n = 193$. At this sample size, the minimum MISE of the local quadratic estimate of Section 5.1.1 matches that of the local con-

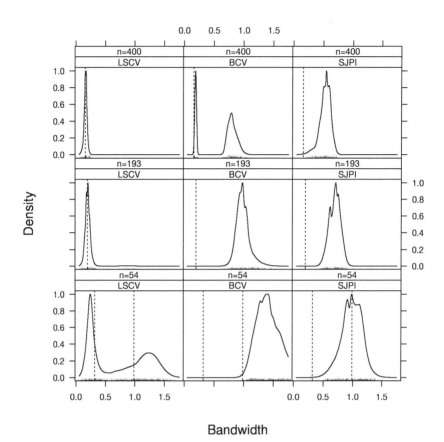

FIGURE 10.6. Selected bandwidths for the claw density. For three sample sizes, three bandwidth selectors are applied to 1000 replications. Density estimates for the selected bandwidths are shown.

stant kernel estimate. For n larger than 193, the performance of the kernel estimate - and hence of the bandwidth selectors - has little relevance, since the local quadratic method provides better estimates. *This is particularly true for plug-in methods, which implicitly use the local quadratic estimate at the pilot stage.*

Figure 10.6 reports some simulation results for the LSCV, BCV and SJPI selectors for the claw density. For three different sample sizes ($n = 54$, $n = 193$ and $n = 400$), 1000 samples of size n are drawn from the claw density. The three bandwidth selectors are applied to each sample. The densities of the selected bandwidths are then plotted.

Clearly, only LSCV displays the behavior that should be expected of a bandwidth selector. At $n = 54$, the distribution of the selected bandwidths displays two modes, centered near the two target values. At $n = 193$, LSCV

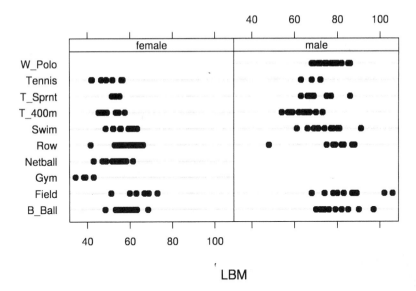

FIGURE 10.7. Australian Institute of Sport dataset. Lean body mass index for 202 athletes.

is fairly reliably finding the claws. At these smaller sample sizes, BCV and SJPI completely fail to find the interesting structure!

10.2.3 Australian Institute of Sport Dataset

What does all this mean for real data? Figure 10.7 shows a dataset consisting of the lean body mass indices of 202 athletes, broken down by gender (female or male) and ten different sports. The data is from Cook and Weisberg (1994).

Figure 10.8 shows two different kernel density estimates for this dataset, similar to Jones, Marron and Sheather (1996). The two bandwidths were selected by LSCV and SJPI. Clearly, the SJPI fit is visually much more pleasing than the LSCV fit.

But at this stage we cannot draw any conclusions as to which fit is best. Figure 10.8 is remarkably similar to Figure 10.6 with the claw density; in light of the simulations in Figure 10.5, it is clear that SJPI will select a large bandwidth without regard to the underlying density.

Jones, Marron and Sheather (1996) note that there are differences between males and females, thus providing an explanation for the flat top or possible bimodality displayed by the SJPI estimate. But differences between sports are also clear from the data in Figure 10.7: The female gymnasts are clearly distinct from the other sports. Athletes in tennis and track events generally have smaller LBM indices than those in rowing or field events.

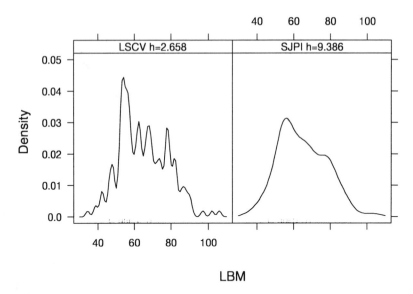

FIGURE 10.8. Kernel density estimates for the AIS dataset. On the left, $h = 2.66$ was selected by LSCV. On the right, $h = 9.39$ was selected by SJPI.

Even within sports, there are differences: the coxswain clearly stands out on both rowing teams. Thus, the dataset is not simply a mixture of male and female distributions, but a mixture of at least 17 different populations, and it is quite possible that some of the other features in the LSCV fit are real.

Figure 10.9 shows the LSCV scores and plug-in relations for the AIS dataset. Again, the LSCV plot immediately provides much information about the relative merits of the smoothing parameters for this dataset. Fits with fewer than seven degrees of freedom are strongly rejected. Although the minimum occurs at about 23 degrees of freedom, fits ranging from 7 to 40 degrees of freedom show relatively little difference in the LSCV scores. For the record, $h = 9.386$ produces a fit with 7.67 degrees of freedom, and the LSCV score is -0.0216.

Once again, the LSCV plot manages to precisely reflect uncertainty in the data. The biggest discrepancy between the two estimates in Figure 10.8 is the height of the peak with the LBM index between 53 and 57. The smaller peaks, while visually less pleasing, have relatively little influence on squared error. To help decide which fit is right, we round the data to the nearest integer. The successive counts for LBM ranging from 53 to 57 are 11, 10, 7, 8 and 9. The expected count, when the density is 0.03135 (the maximum of the SJPI estimate), is $202 \times 0.03135 = 6.33$. Thus the raw counts appear inconsistent with the fitted density, although analyzing

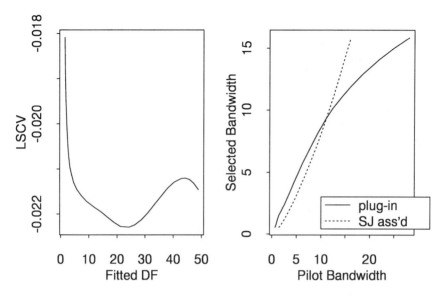

FIGURE 10.9. LSCV scores (left) and plug-in relations (right) for the AIS dataset.

deviance residuals (Exercise 10.4) shows that the discrepancy is not quite enough to declare a significant lack of fit.

The conclusion here is that, once again, the LSCV plot *reflects uncertainty in the data.* Fits smaller than seven degrees of freedom produce a clear lack of fit.

10.3   Conclusions and Further Reading

A number of points have emerged during the course of this chapter:

- The importance of looking at the whole profile of the LSCV (or AIC, or CP) plots, rather than just the minimizer.

- The importance of using the degrees of freedom, rather than the smoothing parameter, as the x-axis for these plots.

- The importance of carefully comparing fits with the data, rather than just looking at the fit. Simply adding data to the fit is a huge help. (Compare Figure 10.3 with figure 6.17 of Scott (1992) or figure 2.2 of Sheather (1992), and look for oversmoothing of the left peak).

- The need to carefully separate prior assumptions built into selectors from the ability to identify features in the data. This is achieved not by looking at single examples, but by looking at pairs of examples and seeing how the selectors respond to changes in the underlying problem.

- Variability of cross validation and classical methods is not a problem. Rather, it is a symptom of the difficulty of bandwidth selection and the problem of resolving uncertainty in the data. Plug-in methods reflect the difficulty by oversmoothing difficult problems and have less ability to resolve uncertainty. *This is particularly true in difficult examples – those with fine features that might be missed just by looking at the data. These are the examples where performance of bandwidth and model selection criteria is most critical.*

The strongest arguments in favor of plug-in methods have been based on asymptotic results, which have not been studied in this chapter. Generally, asymptotic results establish rates of convergence to a true target bandwidth. For classical selection procedures, convergence rates are typically $O_p(n^{-1/10})$ (Rice 1984; Hall and Marron 1987). Plug-in methods achieve rates up to $O_p(n^{-1/2})$ (Hall, Sheather, Jones and Marron 1991). See the references for precise interpretations of these rates.

In the examples presented in this chapter, this kind of asymptotic argument clearly has little relevance. For example, in the claw density, a selector has to choose between two competing models and decide whether or not the claws are sufficiently well supported to represent a real feature. As was argued in Section 10.2.2, the most relevant sample sizes are $54 \le n \le 193$; for larger sample sizes, the claws become easy to detect.

But the problems with the asymptotic analysis are much deeper, as the discussion in Loader (1999) shows. In particular, the results for the rates of convergence make incompatible assumptions. Essentially, the asymptotic results for plug-in methods make extra smoothness assumptions about $f(x)$, to enable the pilot local quadratic methods to work. But then, the plug-in *estimate* is asymptotically beaten by its own pilot *estimate*, making the performance of the *selector* quite uninteresting. Brown, Low and Zhao (1997) and Cleveland and Loader (1996) explicitly show how plug-in based estimates can be beaten: Simply choose a higher order estimate and match bandwidths to equate (or reduce) the variance or fitted degrees of freedom. With this bandwidth matching, the higher order fit has, asymptotically, no bias, so must beat the kernel estimate.

An important related paper is Gu (1998), who studies the statistical relevance of a number of issues related to smoothing parameter selection. In particular, he questions the validity of bandwidth asymptotics and a number of other commonly used measures. He also has several examples relating datasets to selected smoothing parameters.

The examples in this chapter have all used (local constant) density estimates, since this is the setting in which plug-in bandwidth selection methods are developed. But the arguments are equally applicable to local quadratic and higher order methods. This is discussed further in Loader (1999).

10.4   Exercises

10.1 Let $X_1, \ldots, X_n$ be an independent sample from a density $f(x)$, and for any subset A of the real line, let $N(A)$ be the number of observations in A.

a) Show $P(N(A) = 0) = \left(1 - \int_A f(x)dx\right)^n$.

b) Consider a kernel density estimate $\hat{f}(x)$ with a constant bandwidth h and the Gaussian kernel; $W(v) = \exp(-(2.5v)^2/2)$. Consider sets of the form $A = (-\infty, c)$. Show that

$$\int_A \hat{f}(x)dx = \frac{1}{n}\sum_{i=1}^{n} \Phi\left(\frac{2.5(c - X_i)}{h}\right). \tag{10.5}$$

where $\Phi(\cdot)$ is the standard normal distribution function.

c) For the Old Faithful dataset, take $c = 1.67$, so $N(-\infty, c) = 0$. For each of the bandwidths in Table 10.2, compute the integral (10.5) and hence estimate $P(N(A) = 0)$. Which bandwidths are rejected as inadequate?

10.2 Suppose $X(t)$ is a Gaussian process, with mean 0 covariance function $E(X(s)X(t)) = \sigma(s, t)$. The process is observed for $t \in T$ where T is a finite set of points; the object is to predict $X(t_0)$ for some $t_0 \notin T$.

a) A linear predictor has the form $\hat{X}(t_0) = \sum_{t \in T} a_t X(t)$. Using the squared error risk $R(X(t_0), \hat{X}(t_0)) = E(X(t_0) - \hat{X}(t_0))^2$, show that the best linear predictor is

$$\hat{X}(t_0) = \sigma(t_0, T)^T \sigma(T, T)^{-1} X(T).$$

b) Suppose $X(t)$ is white noise, so $\sigma(s, t) = I(s = t)$, and T is the set of integers. Show the best linear predictor is $\hat{X}(t_0) = 0$ for $t_0 \notin T$.

c) Suppose $X(t) = W(t+1) - W(t)$ where $W(t)$ is standard Brownian motion $E(W(s)W(t)) = \min(s, t)$. Show that

$$\sigma(s, t) = E(X(s)X(t)) = (1 - |s - t|)_+,$$

and hence the *observed* $\{X(t), t \in T\}$ is again white noise. Show the best linear predictor is linear interpolation:

$$\hat{X}(t + \lambda) = (1 - \lambda)X(t) + \lambda X(t + 1)$$

where t is an integer and $0 < \lambda < 1$.

Remark: This exercise shows how difficult bandwidth selection can be. The *observed* processes are identical for each model, so no bandwidth selector (or any other data-based statistic) can distinguish the two models. Yet the optimal amount of smoothing changes from infinite to no smoothing!

10.3 In this problem we want to estimate $\int f''(x)^2 dx$ for an unknown density $f(x)$.

  a) Let $g(x) = \log(f(x))$. Show

$$f''(x) = f(x)(g''(x) + g'(x)^2). \tag{10.6}$$

  b) Compute the local log-quadratic fit for the Old Faithful dataset used in Figure 5.1. Compute predictions of $f(x)$, $g'(x)$ and $g''(x)$ on the grid `seq(1,6,length=101)` and hence an estimate of $f''(x)$ using (10.6). Plot your estimate of $f''(x)$.

  c) Using the trapezoidal rule, compute an estimate of $\int f''(x)^2 dx$. Plug this into the optimal bandwidth formula (10.3), when

$$W(v) = e^{-v^2/2}/\sqrt{2\pi};$$

  note that $\int W(v)^2 dv = 1/(2\sqrt{\pi})$ and $\int v^2 W(v) dv = 1$. Compare with the results in Table 10.2.

10.4 Compute a rounded 'count' version of the AIS dataset. Using local Poisson regression, compute local constant fits similar to those in Figure 10.8. Compute and plot the deviance residuals. How strong is the evidence of undersmoothing at $h = 9.39$? Repeat with larger bandwidths.

10.5 Generate a sample of size 100 with

$$X_i = Y_i + Z_i$$

where Y_i is sampled (with replacement) from $10, 20, 30, \ldots, 100$ and $Z_i \sim N(0,1)$. Select bandwidths for estimating the density of the X_i with a Gaussian kernel density estimate, using the LSCV, SJPI, BCV and GKK methods. Plot the density estimates and compare with the data. Also compute cross validation plots and design similar plots for the plug-in methods. Which plots are most informative in relation to the data?

11
Adaptive Parameter Choice

In regression situations with relatively low noise and a large amount of structure, it is quite possible that no single smoothing parameter or degree of local polynomial will provide an adequate fit to the data. In this case, it may be desirable to use *locally adaptive* smoothing methods, which vary the amount of smoothing in a location dependent manner, so as to obtain a satisfactory fit.

The approach taken in this chapter is based on a simple principle:

> In the neighborhood of a fitting point x, does the local polynomial $\langle \hat{a}, A(u - x) \rangle$ fit the data?

In Section 11.1 some measures to assess local goodness of fit are introduced. In Section 11.2, these are applied to locally select the degree of local polynomial and bandwidth. The local goodness of fit approaches were jointly developed with Bill Cleveland and Trevor Hastie; earlier versions of the methods were used in Cleveland and Loader (1996).

Before proceeding, we should add some notes of caution. Locally adaptive procedures work well on examples with plenty of data, obvious structure and low noise. But, as was argued in Chapter 10, these are not the difficult problems for model selection. The real challenges for model selection occur when the structure is not obvious, and there are questions as to which features in a dataset are real. In such cases, simpler methods of bandwidth specification, such as nearest neighbor methods and global cross validation, are most useful, and locally adaptive methods produce little benefit.

It should also be noted that locally adaptive procedures have tuning parameters, or penalties, that control the bias-variance trade-off, and the estimates can be quite sensitive to the choice of these parameters. Thus features should not be declared real simply because they appear in adaptive estimates. Residual analysis and goodness of fit diagnostics are just as important for locally adaptive procedures as they are for global procedures.

Many other approaches to locally adaptive smoothing have been proposed in the literature. The first method based on local polynomial fitting was proposed in Breiman and Meisel (1976). Working in the multidimensional setting, Breiman and Meisel begin with the full linear model. Then, the dataset is split into two equally sized subsets (with a randomly chosen split plane), and linear models fitted to each piece. This process is continued recursively, as long as an F test shows a significant difference between the two subregions. Another algorithm is the supersmoother algorithm of Friedman and Stuetzle (1982), which chooses bandwidths based on a cross validation method. The algorithms of Fan and Gijbels (1995b) (variable bandwidth) and Fan and Gijbels (1995a) combine goodness of fit and plug-in steps.

11.1   Local Goodness of Fit

To develop locally adaptive procedures, we need local analogs of the goodness of fit measures, such as cross validation and CP. Our strategy in deriving a locally adaptive fit is to compute a local goodness of fit for each of a class of candidate fits and thereby select the best local fit. Choice of classes of candidate fits and the fitting of the models are discussed in Section 11.2.

11.1.1   Local CP

For a fixed fitting point x, consider a local polynomial fit using degree p and bandwidth h. Let $\hat{a}$ be the coefficients of the fitted polynomial (computed using (9.8), to allow for nonhomogeneous variance), and let

$$\tilde{\mu}(x_i) = \langle \hat{a}, A(x_i - x) \rangle$$

be the local polynomial evaluated at the data point x_i, The goodness of fit measure is a locally weighted version of the loss (2.22), using the smoothing weights to define the average:

$$L_x(\tilde{\mu}, \mu) = \frac{\sum_{i=1}^n w_i(x)(\tilde{\mu}(x_i) - \mu(x_i))^2/\sigma_i^2}{\sum_{i=1}^n w_i(x)} \tag{11.1}$$

where $\sigma_i^2 = \text{var}(Y_i)$.

We note several points. First, (11.1) uses the value $\tilde{\mu}(x_i)$ of the local polynomial fitted at x and evaluated at x_i, rather than $\hat{\mu}(x_i)$. This considerably simplifies the procedure, since it keeps the estimate $\hat{\mu}(x)$ independent (in a computational, not statistical sense) from the estimate at any other point. Second, the pointwise mean squared errors are weighted by the smoothing weights $w_i(x)$. Third, we divide by the sum of the weights to ensure some comparability between risk values computed at different bandwidths.

The expected value of (11.1) can be decomposed into bias and variance terms:

$$E(L_x(\tilde{\mu}, \mu)) = \frac{1}{\sum_{i=1}^n w_i(x)} \left(\sum_{i=1}^n \frac{w_i(x)}{\sigma_i^2} \tilde{b}(x_i)^2 + \sum_{i=1}^n \frac{w_i(x)}{\sigma_i^2} \text{var}(\tilde{\mu}(x_i)) \right)$$

(11.2)

where $\tilde{b}(x_i) = E(\tilde{\mu}(x_i)) - \mu(x_i)$ is the bias of $\tilde{\mu}(x_i)$. Exercise 11.1 shows that the variance component has the simpler expression

$$\sum_{i=1}^n \frac{w_i(x)}{\sigma_i^2} \text{var}(\tilde{\mu}(x_i)) = \nu(\tilde{\mu})$$

(11.3)

where $\nu(\tilde{\mu})$ is a local analog of the fitted degrees of freedom:

$$\nu(\tilde{\mu}) = \text{tr}\left((\mathbf{X}^T \mathbf{W} \mathbf{V} \mathbf{X})^{-1} \mathbf{X}^T \mathbf{W}^2 \mathbf{V} \mathbf{X} \right).$$

As usual, $\mathbf{X}$ is the design matrix, $\mathbf{W}$ is a diagonal matrix with entries $w_i(x)$ and $\mathbf{V}$ is a diagonal matrix with entries $1/\sigma_i^2$.

A bias-variance decomposition of the local residual sum of squares yields

$$\sum_{i=1}^n \frac{w_i(x)}{\sigma_i^2} E(Y_i - \tilde{\mu}(x_i))^2$$

$$= \sum_{i=1}^n \frac{w_i(x)}{\sigma_i^2} \tilde{b}(x_i)^2 + \sum_{i=1}^n \frac{w_i(x)}{\sigma_i^2} \text{var}(Y_i - \tilde{\mu}(x_i))$$

$$= \sum_{i=1}^n \frac{w_i(x)}{\sigma_i^2} \tilde{b}(x_i)^2 + \text{tr}(\mathbf{W}) - \nu(\tilde{\mu}).$$

(11.4)

Eliminating the unknown bias terms from (11.2) and (11.4) gives the unbiased estimate of the local goodness of fit defined by (11.1). As in the global CP (and perhaps more so, for reasons discussed later), it is useful to also consider a generalized version of the local CP statistic, incorporating an arbitrary penalty on the variance term.

Definition 11.1 The **local generalized CP**, with variance penalty α, is

$$\text{LoCP}_\alpha(\tilde{\mu}) = \frac{1}{\text{tr}(\mathbf{W})} \sum_{i=1}^n \frac{w_i(x)}{\sigma_i^2} (Y_i - \tilde{\mu}(x_i))^2 - 1 + \alpha \frac{\nu(\tilde{\mu})}{\text{tr}(\mathbf{W})}.$$

The case $\alpha = 2$ gives the unbiased estimate of (11.1).

Implementation of the local CP method requires estimating the variances σ_i^2. This can be obtained by a preliminary fit with a small bandwidth and either computing the global estimate (2.18) or using the local variance estimates from Section 9.1.2.

11.1.2 Local Cross Validation

We can develop local versions of the cross validation statistics from Section 2.4.1. First, we require a local analog of the prediction mean squared error (2.19). To assess the performance of $\tilde{\mu}(x)$ within its smoothing window, the goodness of fit measure must be appropriately weighted. This leads to the measure

$$\text{LoPMSE}(\tilde{\mu}) = \frac{1}{EW(\frac{x_{\text{new}} - x}{h})} E\left(W(\frac{x_{\text{new}} - x}{h})(Y_{\text{new}} - \tilde{\mu}(x_{\text{new}}))^2 \right).$$

The local cross validation score estimates this measure using the leave-one-out approach.

Definition 11.2 The **local cross validation** score is

$$\text{LoCV}(\tilde{\mu}) = \frac{\sum_{i=1}^n w_i(x)(Y_i - \tilde{\mu}_{-i}(x_i))^2}{\sum_{i=1}^n w_i(x)}$$

where $\tilde{\mu}_{-i}(x_i)$ is the value of the leave-x_i-out local polynomial.

Theorem 2.2 has a local analog:

$$\text{LoCV}(\tilde{\mu}) = \frac{1}{\sum_{i=1}^n w_i(x)} \sum_{i=1}^n w_i(x)\frac{(Y_i - \tilde{\mu}(x_i))^2}{(1 - \text{infl}(x, x_i))^2}$$

where the local influence function is

$$\text{infl}(x, x_i) = A(x_i - x)^T (\mathbf{X}^T \mathbf{W} \mathbf{X})^{-1} A(x_i - x) w_i(x).$$

Exercise 11.1 shows the locally weighted average of the influence values is

$$\frac{\sum_{i=1}^n w_i(x)\text{infl}(x, x_i)}{\sum_{i=1}^n w_i(x)} = \frac{\nu(\tilde{\mu})}{\text{tr}(\mathbf{W})}. \tag{11.5}$$

This enables us to define a localized version of generalized cross validation.

Definition 11.3 The **local generalized cross validation** statistic is defined as

$$\text{LoGCV}(\hat{\mu}) = \text{tr}(\mathbf{W})\frac{\sum_{i=1}^n (Y_i - \tilde{\mu}(x_i))^2}{(\text{tr}(\mathbf{W}) - \nu(\tilde{\mu}))^2}.$$

11.1.3   Intersection of Confidence Intervals

The intersection of confidence intervals (ICI), introduced in Goldenshluger and Nemirovski (1997) and further developed by Katkovnik (1998) and Katkovnik (1999) provides an alternative method of assessing local goodness of fit.

For a local regression estimate $\hat{\mu}_h(x)$, computed with a small bandwidth h, a confidence interval for the mean $\mu(x)$ has the form

$$I_h(x) = (\hat{\mu}_h(x) - c\hat{\sigma}\|l_h(x)\|, \hat{\mu}_h(x) + c\hat{\sigma}\|l_h(x)\|).$$

As the bandwidth h is increased, the standard deviation of $\hat{\mu}_h(x)$, and hence $\|l_h(x)\|$, decreases. Thus, the confidence intervals become narrower. If h is increased too far, the estimate $\hat{\mu}_h(x)$ will become heavily biased. Eventually, the confidence intervals will become inconsistent, in the sense that the intervals constructed at different bandwidths have no common intersection.

The ICI bandwidth is then defined to be the largest bandwidth h_1 such that

$$\bigcap_{h_0 \leq h \leq h_1} I_h(x)$$

is nonempty.

The main tuning parameter in this procedure is the critical value c in constructing the confidence intervals. Choosing $c = 1.96$ leads to 95% confidence intervals under normality assumptions. But the resulting adaptive estimates tend to display bias. Smaller values of c, for example, $c = 1$, sometimes produce visually superior accuracy. The bias-variance trade-off on c and selection of c through cross validation are studied in Katkovnik (1998, 1999).

The ICI method can be used in conjunction with any estimate for which approximate standard errors are available. For example, it extends immediately to local likelihood or local slope parameters. Applications with nonlinear observations can be found in Katkovnik (1996) and Katkovnik and Stankovic (1998).

A procedure closely related to ICI was introduced in Lepski, Mammen and Spokoiny (1997). This method uses the standard deviation of the differences $\hat{\mu}_{h_1}(x) - \hat{\mu}_h(x); h \leq h_1$ and increases h_1 until a significant difference is found.

11.1.4   Local Likelihood

The local goodness of fit criteria have been derived for local regression models, but they can easily be extended to local likelihood. A localized version of the likelihood cross validation statistic (4.9) is

$$\frac{1}{\mathrm{tr}(\mathbf{W})} \sum_{i=1}^{n} w_i(x) D(Y_i, \tilde{\theta}_{-i}(x_i))$$
(11.6)

where $D(y, \theta)$ is the deviance function and $\tilde{\theta}_{-i}(x_i) = \langle \hat{a}_{-i}, A(x_i - x) \rangle$ where $\hat{a}_{-i}$ is the leave-x_i-out estimate of the local parameters. Using a one-term linearization of $\hat{a}_{-i}$, similar to that in Exercise 4.6, yields

$$\tilde{\theta}_{-i}(x_i) \approx \tilde{\theta}(x_i) - A(x_i - x)^T (\mathbf{X}^T \mathbf{W} \mathbf{V} \mathbf{X})^{-1} A(x_i - x) w_i(x) \dot{l}(Y_i, \tilde{\theta}(x_i))$$

and

$$D(Y_i, \tilde{\theta}_{-i}(x_i)) \approx D(Y_i, \tilde{\theta}(x_i))$$
$$-2A(x_i - x)^T (\mathbf{X}^T \mathbf{W} \mathbf{V} \mathbf{X})^{-1} A(x_i - x) w_i(x) \dot{l}(Y_i, \tilde{\theta}(x_i))^2.$$

Substituting this approximation into (11.6) and replacing $\dot{l}(Y_i, \tilde{\theta}(x_i))^2$ by $-\ddot{l}(Y_i, \tilde{\theta}(x_i))$, leads to the local Akaike criterion. We also allow an arbitrary penalty parameter α.

Definition 11.4 The **local generalized Akaike information criterion**, with penalty parameter α, is

$$\mathrm{LoAIC}_\alpha(\tilde{\theta}) = \frac{\sum_{i=1}^{n} w_i(x) D(Y_i, \tilde{\theta}_i)}{\mathrm{tr}(\mathbf{W})} + \alpha \frac{\nu(\tilde{\theta})}{\mathrm{tr}(\mathbf{W})}.$$

11.2 Fitting Locally Adaptive Models

Locally adaptive fits are obtained by computing a number of candidate fits and using one of the criteria introduced in the previous section to select among the fits. LOCFIT supports two classes of candidate fits, obtained either by varying the degree of the local polynomial or by varying the bandwidth. The latter is necessary where a high amount of adaptivity is required, but can be less reliable in situations with high noise.

Example 11.1. A variable degree fit is obtained by specifying a `deg` argument with two components. For example, `deg=c(0,3)` selects among local constant, linear, quadratic and cubic fits. We apply this to the ethanol dataset using a 30% nearest neighbor bandwidth:

```
> fit <- locfit(NOx~E,data=ethanol,alpha=0.3,deg=c(0,3))
> plot(fit,get.data=T)
> plot(predict(fit,what="deg"),type="p",ylab="degree")
```

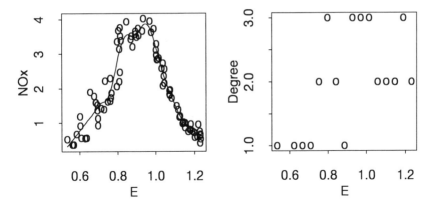

FIGURE 11.1. Variable order fit for ethanol dataset (left) and selected degrees (right).

Figure 11.1 shows the fit. The left panel shows the variable order fit. In the right panel, the selected degrees for each fitting point are shown: `deg=1` (local linear) is selected on the left, where the data is fairly linear and noise is largest. Around the peak and on the right, local quadratic and cubic fits are mostly preferred.

By default, the variable order fit is selected by the localized GCV criterion (11.3). As in the global case, this has the advantage of not requiring a variance estimate. If the local CP criterion is required, this can be requested by providing a third penalty component to the smoothing parameter `alpha`.

The LOCFIT implementation of a local variable bandwidth smoother is more sophisticated. The algorithm aims to keep the number of bandwidths tried as small as possible, to ensure that the computations can be performed in a reasonable time. In particular, it is important to avoid computing unnecessary fits with large bandwidths.

The implementation proceeds in three steps:

1. Choose an initial bandwidth h_0 to cover $p + 2$ nearest neighbors, where p is the number of parameters in the local model. This is close to the smallest bandwidth for which the estimate is well defined.

2. Increase the bandwidth by a factor of 1.3, or in more than one dimension, $1 + 0.3/d$. Repeat until the goodness-of-fit criterion (local CP, AIC or cross validation) shows a sharp increase. The present implementation requires increases of the criterion on at least three successive steps, and the criterion to be at least 50% above its minimum.

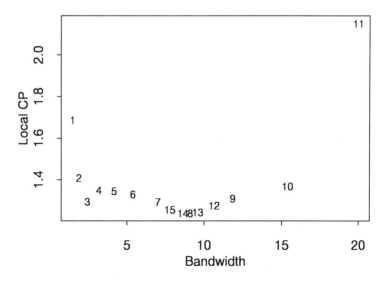

FIGURE 11.2. Local bandwidth selection. Beginning at a small bandwidth (point 1), the bandwidth is increased in large steps until there is a large increase in the estimated risk at point 11. Then a finer search is performed around the minimum (points 12 to 15).

3. Beginning at the bandwidth with the minimum goodness of fit value observed in step 2, perform a finer search for the final bandwidth, with a search factor of $1 + 0.1/d$.

The implementation of the ICI method is similar, but the second step is terminated as soon as the intersection of the confidence intervals is empty.

Example 11.2. The motorcycle dataset used in Example 9.1 could clearly benefit from a local bandwidth, since the dataset has both flat and sharply curved regions. Figure 11.2 shows the local bandwidth selection for a single fitting point; in this case at `time=33.45`. This uses the local variance estimate derived in Example 9.1 and the local CP criterion.

The first fit is computed with $h = 1.45$; this is sufficient to get a well-defined estimate. As the bandwidth is increased, a local minimum (point number 3) is observed in Figure 11.2. But the increase that follows is not sharp, so the search continues. A second local minimum (point number 8) is observed. The following points (9, 10 and 11) exhibit a sharp increase in the local CP criterion, indicating lack of fit of the estimate. The procedure concludes with a finer search (points 12, 13, 14 and 15) around point 8.

Performing this procedure at a large number of points would be computationally expensive for large datasets. Also, the bandwidth as defined may be a discontinuous function of x, leading to a discontinuous fit. However,

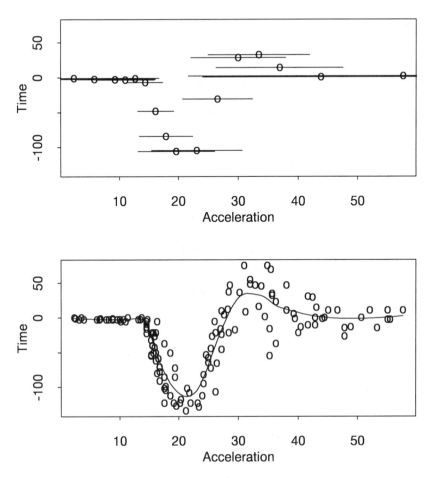

FIGURE 11.3. Fitting points and bandwidths selected for the motorcycle dataset by the adaptive fitting algorithm (top) and the resulting fit (bottom).

the computational methods of LOCFIT come to our aid: By using adaptive fitting structures, the fit will only be performed at sufficient points to define a reasonable estimate. In regions where large bandwidths are sufficient, fitting points will be sparse. Where smaller bandwidths are necessary, there will be more fitting points, but evaluation is relatively cheap in this case. The prediction methods then interpolate between the fitting points, resulting in a continuous fit.

Example 11.3. We continue with the motorcycle dataset. A locally adaptive bandwidth selection is specified using the `acri` arugument. For example, `acri="cp"` uses the local CP rule. A third component to the smoothing parameter `alpha` specifies the variance penalty in (11.1). It is also crucial to include a variance estimate (which may be either local or global, as appropriate to the dataset). Here, we use the local variance estimate from Example 9.1:

```
> fit <- locfit(accel~time, alpha = c(0,0,2), weights=1/vp,
+    acri="cp", data=mcyc.n)
> plot(accel~time, data=mcyc, type="n")
> x <- knots(fit, what="x")
> coef <- predict(fit, where="fitp")
> h <- predict(fit, where="fitp", what="band")
> points(x, coef)
> segments(x-h, coef, x+h, coef)
> plot(fit, get.data=T)
```

Figure 11.3 shows the resulting fit. The top panel shows the selected fitting points (in this case, 16 points) and the widths of the smoothing windows. The smoothing points are closest together at the point of impact where the smallest bandwidths are selected, and furthest apart at later times where the largest bandwidths are used.

Example 11.4. Following Donoho and Johnstone (1994) we define the Dopler function

$$\mu(x) = 20\sqrt{x(1-x)}\sin(2\pi\frac{1.05}{x+0.05}).$$

The design consists of 2048 data points, equally spaced on $[0, 1]$. The errors are independent standard normal random variables. Figure 11.4 shows the generated dataset:

```
> x <- seq(0, 1, length=2048)
> y <- 20*sqrt(x*(1-x))*sin((2*pi*1.05)/(x+0.05))+rnorm(2048)
> plot(x, y, pch=".")
```

We use the local generalized CP statistic, with penalty $\alpha = 4$:

```
> fit <- locfit(y~x, alpha=c(0,0,4), maxk=500, acri="cp")
> plot(x, y)
> plot(fit, mpv=2048)
> plot(preplot(fit, what="band", where="fitp"),
+    type="p", log="y", ylab="bandwidth")
```

The resulting fit is shown in the middle panel of Figure 11.4. The loss $\sum_{i=1}^{n}(\hat{\mu}(x_i) - \mu(x_i))^2)$ is 148.0. Fairly similar results were obtained for values of α between 2.5 and 5. For smaller values of α, spurious features begin to appear in the fit, while for larger values of α, bias begins to show.

The bottom panel of Figure 11.4 shows the selected fitting points and bandwidths. In this example, 124 fitting points were selected. Most of these

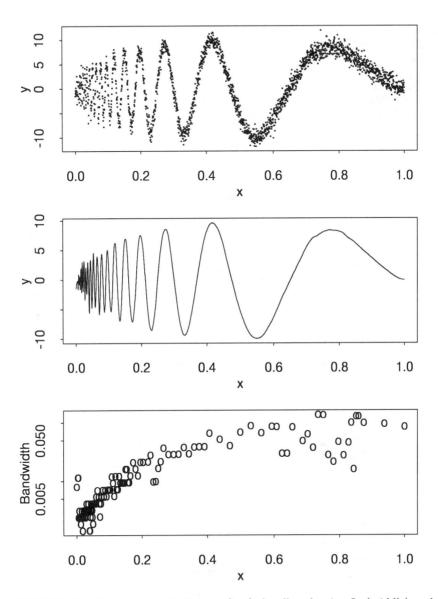

FIGURE 11.4. Dopler example dataset (top), locally adpative fit (middle) and fitting points and bandwidths (bottom).

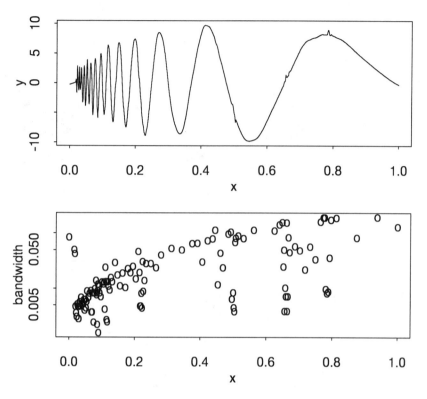

FIGURE 11.5. Fitting the Dopler dataset with the ICI criterion.

are clustered near the left boundary where the smallest bandwidths are used.

Example 11.5. We repeat the Dopler example, using the ICI criterion. This is specified with `acri="ici"`. The critical value c in the confidence intervals is again specified as the third component of the smoothing parameter argument `alpha`. In this example, we use $c = 1.1$:

```
> fit <- locfit.raw(x, y, alpha=c(0,0,1.1),
+    acri="ici", maxk=500)
> plot(fit, mpv=2048)
> plot(preplot(fit, what="band", where="fitp"),
+    type="p", log="y", ylab="bandwidth")
```

Figure 11.5 shows the resulting fits and selected bandwidths. In this case, the loss was 379.6. Smaller values of c produced lower loss (the minimum was 301.6, at $c = 0.9$), although the resulting estimates were less satisfying visually. The choice of c seems to be quite delicate in this example; some spurious noise is already beginning to appear in Figure 11.5.

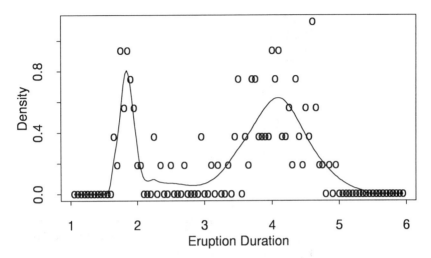

FIGURE 11.6. Locally adaptive fit to the Old Faithful geyser dataset.

Our final example fits a locally adaptive model using local likelihood. In the analysis of the Old Faithful dataset in Chapters 5 and 10, we struggled to find a fit that was satisfactory for both peaks simultaneously. This suggests that locally adaptive methods may be useful.

One could apply the locally adaptive setting directly in the density estimation setting. But rounding and ties in the data, which caused some difficulty for global bandwidth selection, cause even more difficulty for locally adaptive rules (should three tied observations represent a peak in the data and require a small bandwidth?). Visually, the most satisfying results are obtained using the rounded versions of the data (Example 5.8), and using local Poisson regression.

Example 11.6. A locally adaptive Poisson regression is fitted to the rounded version of the Old Faithful dataset:

```
> fit <- locfit(count~duration, weights=rep(107*0.05,99),
+    data=geyser.round, alpha=c(0,0,2), family="poisson")
> plot(fit, get.data=T, mpv=200)
```

Note the weights argument; this ensures that the resulting fit is correctly scaled to estimate the density.

Figure 11.6 shows the resulting fit. Both peaks are smoothly reproduced, and the left peak is sharp, as the data suggests it should be.

11.3   Exercises

11.1   a) In the notation of Section 11.1.1, show that

$$\begin{pmatrix} \tilde{\mu}(x_1) \\ \vdots \\ \tilde{\mu}(x_n) \end{pmatrix} = \mathbf{X}^T \left(\mathbf{X}^T \mathbf{W} \mathbf{V} \mathbf{X} \right)^{-1} \mathbf{X}^T \mathbf{W} \mathbf{V} Y.$$

Thus, prove (11.3).

b) Show that

$$\sum_{i=1}^{n} \frac{w_i(x)}{\sigma_i^2} \mathrm{cov}(Y_i, \tilde{\mu}(x_i)) = \nu(\tilde{\mu}),$$

and hence establish (11.4).

c) Prove (11.5).

11.2 Generate a dataset from the Dopler model (Example 11.4). Fit the locally adaptive models from (11.4) and (11.5), but with different values of the adaptive penalty. Which penalties produce the visually best fits? Look both at the smooth parts of the estimate on the right and the rough parts near $x = 0$.

11.3 Repeat the previous exercise for the other examples from Donoho and Johnstone (1994). Try both local quadratic and local linear fits.

12
Computational Methods

Computational algorithms for local likelihood estimation can be divided into several steps. First, one has to select points x at which to compute the fit. Second, one needs algorithms to compute the fit at these points. Third, one needs to compute diagnostic information, such as variances and the influence function.

The second step is a fairly routine problem, requiring numerical solution of the local likelihood equations. An approach is discussed in Section 12.1.

The first step is more interesting. The main computational ideas underlying LOCFIT build on those incorporated in LOESS (Grosse 1989; Cleveland and Grosse 1991). The local fitting algorithm is carried out on a set of vertices in the design space, and certain interpolants are used to construct function estimates over the cells. However, the LOCFIT implementation differs from LOESS in several respects. In particular, new fitting structures use the bandwidth, rather than the density of design points, to decide when cells are split. The algorithms are described fully in Section 12.2.

Section 12.3 discusses the influence and variance functions. Section 12.4 discusses some issues specific to the density estimation problem.

12.1   Local Fitting at a Point

To find the local likelihood estimate at a single point, we must numerically solve the local likelihood equations (4.14). Since there is no closed form solution, an iterative procedure must be used. The multivariate generalization of the Newton-Raphson method (Conte and De Boor, 1980, section 5.2) provides a convenient procedure.

Suppose we have a guess $\hat{a}^{(k)}$ of the solution of the local likelihood equations (4.15). A linear expansion of the equations around the parameter vector $\hat{a}^{(k)}$ gives

$$\mathbf{X}^T \mathbf{W} \dot{l}(Y, \mathbf{X}(\hat{a}^{(k)} + \nabla a)) \approx \mathbf{X}^T \mathbf{W} \dot{l}(Y, \mathbf{X}\hat{a}^{(k)}) - \mathbf{X}^T \mathbf{W} \mathbf{V} \mathbf{X} \nabla a, \quad (12.1)$$

where $\mathbf{V}$ has diagonal entries $-\ddot{l}(Y_i, \langle \hat{a}^{(k)}, A(x_i - x) \rangle)$. Taking

$$\nabla a^{(k)} = (\mathbf{X}^T \mathbf{W} \mathbf{V} \mathbf{X})^{-1} \mathbf{X}^T \mathbf{W} \dot{l}(Y, \mathbf{X}\hat{a}^{(k)}),$$

the right-hand side of (12.1) is easily seen to equal 0. This leads to the Newton-Raphson iteration:

$$a^{(k+1)} = a^{(k)} + (\mathbf{X}^T \mathbf{W} \mathbf{V} \mathbf{X})^{-1} \mathbf{X}^T \mathbf{W} \dot{l}(Y, \mathbf{X}\hat{a}^{(k)}). \quad (12.2)$$

Any method for solving linear systems of equations can be used to find the vector $\hat{a}^{(k+1)}$. The method presently used by LOCFIT is as follows: Let $\mathbf{D}$ be the matrix of diagonal elements of $\mathbf{J} = \mathbf{X}^T \mathbf{W} \mathbf{V} \mathbf{X}$, and let $\mathbf{P}$ and $\mathbf{\Sigma}$ be the eigenvectors and eigenvalues of $\mathbf{D}^{-1/2} \mathbf{J} \mathbf{D}^{-1/2}$, so that

$$\mathbf{J} = \mathbf{D}^{1/2} \mathbf{P}^T \mathbf{\Sigma} \mathbf{P} \mathbf{D}^{1/2}.$$

The system of equations (12.2) can then be written

$$a^{(k+1)} = a^{(k)} + \mathbf{D}^{-1/2} \mathbf{P}^T \mathbf{\Sigma}^{-1} \mathbf{P} \mathbf{D}^{-1/2} \mathbf{X}^T \mathbf{W} \dot{l}(Y, \mathbf{X}\hat{a}^{(k)}).$$

This can easily be evaluated in a right-to-left manner.

To complete the implementation, some additional questions must be addressed:

1. What is the starting value $\hat{a}^{(0)}$?

2. Does the algorithm converge?

3. If the algorithm converges, does it converge to the maximizer of the local likelihood $\mathcal{L}_x(a)$?

4. What happens if the Jacobian matrix $\mathbf{J} = \mathbf{X}^T \mathbf{W} \mathbf{V} \mathbf{X}$ is singular?

For most of the models implemented in LOCFIT the local constant estimate has a closed form expression. For likelihood regression models,

$$\hat{a}_0 = g(\bar{Y})$$

where $g(\cdot)$ is the link function and $\bar{Y}$ is the locally weighted sample average. This provides a convenient starting value for the Newton-Raphson iteration; local slopes and higher order coefficients are set to zero.

The local constant starting values are usually poor approximations to the final coefficient estimates, and one may be tempted to try more complicated approaches. For example, one might be tempted to try using local regression to find starting values for all coefficients. But in this author's experience, this saves at most one iteration, and has the added complexity of programming the initial step. Thus the benefits would appear minimal. Another temptation is to use coefficients from the estimate at one fitting point to provide starting values at a neighboring point. But again this doesn't save much, unless one is fitting on a very fine grid of points.

Does the Newton-Raphson algorithm converge? First, consider how the local likelihood changes in the direction the Newton-Raphson iteration moves. Provided the matrix $\mathbf{J}$ is positive definite (concavity of the log-likelihood ensures this; see Section 4.4), Exercise 12.1 shows the vector $\nabla a^{(k)}$ is an ascent direction for the local likelihood:

$$\mathcal{L}_x(a^k + \lambda \nabla a^{(k)}) \geq \mathcal{L}_x(a^k)$$

for small positive λ. However, this doesn't necessarily hold for $\lambda = 1$, and a direct implementation of the Newton-Raphson method may not converge.

Instead, we consider a damped Newton-Raphson algorithm, choosing

$$a^{(k+1)} = a^{(k)} + \frac{1}{2^j} \mathbf{J}^{-1} \mathbf{X}^T \mathbf{W} i.$$

Here, j is selected at each step to be the smallest non-negative integer that results in an increase of the local log-likelihood. With this modification, the local likelihood is guaranteed to increase at each step and converge to a local maximum. For concave likelihoods, this local maximum must also be the global maximum.

In Section 4.4 and elsewhere, conditions were derived for existence and uniqueness of the local likelihood estimate under various models. When these conditions fail, various instabilities occur in the Newton-Raphson algorithm; for example, the Jacobian matrix $\mathbf{J}$ may be singular, or the parameters may diverge to ∞. The conditions of Theorem 4.1 can be difficult to check directly, so checks for divergence are implemented as part of the Newton-Raphson algorithm.

12.2   Evaluation Structures

As mentioned in Section 3.1 and elsewhere, LOCFIT does not (by default) perform local regression directly at every point. Rather, it selects a set of evaluation points at which the fit is performed directly, then uses the information from the fits at these points to interpolate elsewhere.

This section describes the manner in which evaluation points are selected, and the interpolation schemes used. The goal here is efficiency: since the interpolation problem is much cheaper than the local fitting problem, the number of evaluation points should be kept as small as possible, but retaining sufficient information to approximate the entire local regression surface.

12.2.1 Growing Adaptive Trees

The simplest evaluation structure is a grid of points. Fitting limits are determined for each predictor variable (usually the range of the variable) and each side is divided by a specified number of grid lines. The local fit is then carried out at each vertex of the grid.

But this is inefficient, particularly when nearest neighbor or locally adaptive bandwidth schemes are used. In these situations, the fitted surface may be very smooth in regions where large bandwidths are used but rough in regions where small bandwidths are used. Ideally, more evaluation points should be placed in the small bandwidth regions.

The k-d tree structure (Friedman, Bentley and Finkel 1977) was used for this purpose in LOESS (Cleveland and Grosse 1991). One begins by bounding the data in a rectangular box and evaluating the fit at the vertices of the box. One then recursively splits the box into two pieces, then each subbox into two pieces, and so on. At each stage, the split is chosen so that the remaining data points are divided into two subsets of equal size. Refinement continues until each lowest level box contains at most k points, where

$$k = 0.2n\alpha.$$

Here α is the nearest neighbor component of the bandwidth.

The main evaluation structure used in LOCFIT is a tree-based structure similar to the quadtree in approximation theory literature (Seidel 1997). The algorithm begins by bounding the dataset in a rectangle and successively dividing into two equally sized pieces. While based on similar principles to the k-d tree, this has a number of important differences:

- Recursion is based on bandwidths rather than on number of design points. Thus the algorithm isn't restricted to nearest neighbor bandwidths.

- The decision to split an edge is based solely on the bandwidths at the two ends of the edge; the rest of the cell is irrelevant.

- An edge is always split at the midpoint.

The first point is important, since it means the algorithm selects points according to the resolution of the estimate, and is no longer restricted to nearest neighbor bandwidths. For example, Figure 11.4 showed the fitting points selected for the Dopler example. This varies in accordance with the

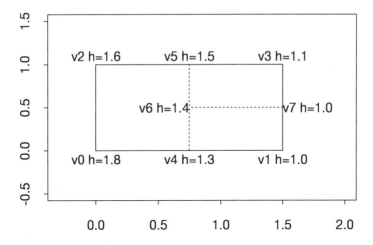

FIGURE 12.1. Growing a rectangular tree. The cell (v_0, v_1, v_2, v_3) is split by the line (v_4, v_5). Then, (v_4, v_1, v_5, v_3) is split by (v_6, v_7).

locally adaptive bandwidth: Most fitting points are in regions where small bandwidths are chosen.

The second and third points are important at the interpolation stage and for improved efficiency. In two (or more) dimensions, an edge may be shared by two (or more) adjacent cells. Under the k-d tree algorithm, the edge may be split in two different places, and preserving continuity of the surface becomes a difficult problem.

The procedure by which LOCFIT grows an adaptive tree is illustrated in Figure 12.1. Initially, the dataset is bounded in a box, and local fitting is carried out at the vertices. For an edge of the box joining vertices v_i and v_j, a score is defined:

$$\rho_{i,j} = \frac{\|v_i - v_j\|}{\min(h_i, h_j)} \tag{12.3}$$

where h_i and h_j are the bandwidths used at the vertices. Any edge whose score exceeds a critical value c ($c = 0.8$ by default) needs to be split.

In Figure 12.1, the initial rectangle has vertices at (v_0, v_1, v_2, v_3), and suppose the local fitting algorithm selected bandwidths $(1.8, 1.0, 1.6, 1.1)$. For the present discussion, it doesn't matter how these bandwidths are obtained. The generation of evaluation structures depends only on the bandwidths and not the algorithm used to generate them.

The longest edges of the initial rectangle are (v_0, v_1) and (v_2, v_3); the scores are $\rho_{0,1} = 1.5/1.0 = 1.5$ and $\rho_{2,3} = 1.5/1.1 = 1.36$ respectively. Both these sides require splitting, so new vertices are added at $v_4 = (0.75, 0)$ and $v_5 = (0.75, 1.0)$. Suppose the bandwidths at these points are $h_4 = 1.3$ and

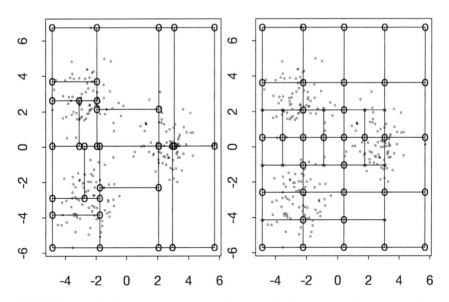

FIGURE 12.2. Evaluation structures: A k-d tree (left) and adaptive rectangular tree (right). In both cases, 34 evaluation points result.

$h_5 = 1.5$. The left cell (v_0, v_4, v_2, v_5) does not require any further splits, since the largest score is $\rho_{4,5} = 1.0/1.3 = 0.77$.

The right cell (v_4, v_1, v_5, v_3) does require splitting, since $\rho_{1,3} = 1.0/1.0 = 1.0$. We now have a problem. The horizontal split requires adding two vertices, v_6 and v_7. But v_6 splits the edge (v_4, v_5), and the score (12.3) dictates that this edge should not be split. The local fit at this vertex can't be used, since this would lead to the k-d tree blending problem.

The solution is to create v_6 as a pseudo-vertex. Rather than carrying out the split directly at v_6, the fit will be defined from the fits at v_4 and v_5 in a manner so as to preserve continuity and differentiability across the cells.

Example 12.1. Figure 12.2 shows a k-d tree (left) and adaptive rectangular tree (right) for the trimodal dataset used in example 5.5. In both cases a nearest neighbor smoothing parameter $\alpha = 0.35$ is used, and fitting is performed directly at 34 points (for the rectangular tree, this required slight adjustment of the default cut parameter). The points selected by the rectangular tree algorithm cover the data more efficiently; in several places, the k-d tree algorithm chooses two points very close to each other.

Note the plot.eval() function is used to plot the evaluation structure:

```
> fit <- locfit(~x0+x1, data=trimod, deg=1,
+   alpha=0.35, ev="kdtree")
> plot.eval(fit)
> points(trimod$x0, trimod$x1, cex = 0.3)
```

```
> fit <- locfit(~x0+x1, data=trimod, deg=1,
+    alpha=0.35, cut=0.85)
> plot.eval(fit)
> points(trimod$x0, trimod$x1, cex = 0.3)
```

12.2.2 Interpolation Methods

After computing the fit at the evaluation points, we need to specify an interpolation method to define the fit at any other point. For simplicity, we first consider the one dimensional case.

The simplest method is linear interpolation. If $v_0 \leq x \leq v_1$ for two evaluation points v_0 and v_1, we define

$$\hat{\mu}(x) = \hat{\mu}(v_0)\frac{x - v_1}{v_0 - v_1} + \hat{\mu}(v_1)\frac{x - v_0}{v_1 - v_0}.$$

Linear interpolation is usually unsatisfactory, since the resulting curve is not differentiable. A fine grid of points will be needed to avoid trimming in regions of high curvature.

To reduce these problems, the LOCFIT algorithm also uses derivative estimates at the vertices; these are readily available when local linear or higher order fitting is used. To interpolate over the interval $[v_0, v_1]$, the four values $\hat{\mu}(v_0)$, $\hat{\mu}(v_1)$, $\hat{\mu}'(v_0)$ and $\hat{\mu}'(v_1)$ are used. These define a unique cubic polynomial, given explicitly by

$$
\begin{aligned}
\hat{\mu}(x) &= \phi_0(\lambda)\hat{\mu}(v_0) + \phi_1(\lambda)\hat{\mu}(v_1) \\
&\quad + (v_1 - v_0)\left(\psi_0(\lambda)\hat{\mu}'(v_0) + \psi_1(\lambda)\hat{\mu}'(v_1)\right)
\end{aligned}
\tag{12.4}
$$

where

$$
\begin{aligned}
\lambda &= \frac{x - v_0}{v_1 - v_0} \\
\phi_0(\lambda) &= (1 - \lambda)^2(1 + 2\lambda) \\
\phi_1(\lambda) &= \lambda^2(3 - 2\lambda) \\
\psi_0(\lambda) &= \lambda(1 - \lambda)^2 \\
\psi_1(\lambda) &= -\lambda^2(1 - \lambda).
\end{aligned}
$$

Some important properties of this interpolation scheme are:

1. The resulting surface has a globally continuous first derivative, since the fitted derivatives are preserved at the vertices v_i.

2. Preservation of a polynomial up to degree 3. This is important, since the attractive properties of local polynomial fitting stem from its local nature and preservation of polynomials up to the degree of fit. The interpolation scheme preserves this for fitting up to local cubic.

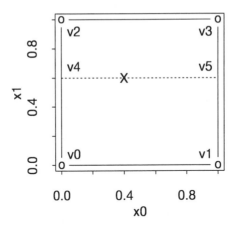

FIGURE 12.3. Interpolation over a grid cell. First, collapse the cell in the vertical direction, then in the horizontal direction.

3. Rapid to compute at a large number of points. This contrasts with direct fitting, which requires going back to the original data.

This construction can be extended to the multivariate case by successively collapsing each dimension, as illustrated in Figure 12.3. Suppose a point $x = (x_0, x_1)$ is in the cell bounded by vertices $[v_0, v_1, v_2, v_3]$. Define

$$\lambda_i = \frac{x_i - v_{0,i}}{v_{3,i} - v_{0,i}}; i = 0, 1.$$

First, the rectangle is collapsed along the vertical axis, interpolating the function values and horizontal derivatives to v_4 and v_5. This yields

$$\hat{\mu}(v_4) = \phi_0(\lambda_1)\hat{\mu}(v_0) + \phi_1(\lambda_1)\hat{\mu}(v_2) + \psi_0(\lambda_1)\hat{\mu}_1(v_0) + \psi_1(\lambda_1)\hat{\mu}_1(v_1)$$

and

$$\hat{\mu}_0(v_4) = \phi_0(\lambda_1)\hat{\mu}_0(v_0) + \phi_1(\lambda_1)\hat{\mu}_0(v_2) + \psi_0(\lambda_1)\hat{\mu}_{0,1}(v_0) + \psi_1(\lambda_1)\hat{\mu}_{0,1}(v_1)$$

and similar expressions for $\hat{\mu}(v_5)$ and $\hat{\mu}_0(v_5)$. The subscripts of $\hat{\mu}$ denote derivatives: $\hat{\mu}_1$ is derivative in the vertical direction, $\hat{\mu}_0$ in the horizontal direction and $\hat{\mu}_{0,1}$ is the mixed second order derivative. Secondly, the function values are interpolated to the point x, along the dashed line:

$$\hat{\mu}(x) = \phi_0(\lambda_0)\hat{\mu}(v_4) + \phi_1(\lambda_0)\hat{\mu}(v_5) + \psi_0(\lambda_0)\hat{\mu}_0(v_4) + \psi_1(\lambda_0)\hat{\mu}_0(v_5)$$

The remaining problem is pseudo-vertices, such as v_6 in Figure 12.1, where the fit is not computed directly. For computation of the interpolant over the cell v_6, v_7, v_5, v_3, the value $\hat{f}(v_6)$ is defined as the cubic interpolant along v_4, v_5. This preserves continuity on the adjacent cells. Formally, we have

$$\hat{\mu}(v_6) = \frac{1}{2}(\phi_0(0.5)\hat{\mu}(v_4) + \hat{\mu}(v_5)) + \frac{1}{8}(\hat{\mu}_{0,1}(v_4) - \hat{\mu}_{0,1}(v_5))$$

$$\hat{\mu}_{1,0}(v_6) = \frac{1}{2}(\hat{\mu}_{1,0}(v_4) + \hat{\mu}_{1,0}(v_5)) + \frac{1}{8}(\hat{\mu}_{1,1}(v_4) - \hat{\mu}_{1,1}(v_5))$$

$$\hat{\mu}_{0,1}(v_6) = -\frac{3}{2}(\hat{\mu}(v_4) - \hat{\mu}(v_5)) - \frac{1}{4}(\hat{\mu}_{0,1}(v_4) + \hat{\mu}_{0,1}(v_5))$$

$$\hat{\mu}_{1,1}(v_6) = -\frac{3}{2}(\hat{\mu}_{1,0}(v_4) - \hat{\mu}_{1,0}(v_5)) - \frac{1}{4}(\hat{\mu}_{1,1}(v_4) + \hat{\mu}_{1,1}(v_5)).$$

It is important to interpolate *all* coefficients at the pseudo-vertices. If a local linear model is fitted, then by construction we use $\hat{\mu}_{1,1}(v_4) = \hat{\mu}_{1,1}(v_5) = 0$. But the interpolated value $\hat{\mu}_{1,1}(v_6)$ defined by (12.5) will usually be nonzero.

12.2.3   Evaluation Structures in LOCFIT

By default, LOCFIT uses the adaptive rectangular tree structures described earlier. But a number of other structures are also supported and can be specified by the ev argument:

- ev="grid" for a grid.

- ev="kdtree" for a k-d tree.

- ev="tree" for an adaptive rectangular tree.

- ev="phull" for triangulation, similar to Loader (1994).

- ev="data" for direct evaluation at data points.

- ev="cross" for leave-one-out cross validation at data points.

The default is the tree structure, described in the previous section. The phull structure bounds the data in a polyhedron and divides the polyhedron into triangles; the division is again based on an adaptive scheme using bandwidths. The interpolant is constructed using the Clough-Tocher method (Clough and Tocher 1965).

The data and crossval structures evaluate directly at each data point; crossval uses leave-one-out cross validation. Due to computational cost, this should only be used for n small. Note also that these structures can't be interpolated, and the fits can't be plotted.

Several other parameters also control the evaluation structure. For the rectangular structures, flim enables the user to specify the initial bounding box for the fit.

For grid structures, `mg` is used to control the number of grid lines per side. The default is 10. In the multivariate case, `mg` may be either a single number or a vector.

For the rectangular tree and triangulation structures, the `cut` specifies the critical value used in the score criterion (12.3).

An important argument is `maxk`. For the adaptive structures, it is difficult to guess in advance how many vertices the final tree will contain. Thus, LOCFIT can't always guess the right amount of memory to allocate in advance, and dynamic allocation isn't feasible in the S interface. If the `locfit()` call produces a warning about insufficient space, then increase `maxk`. This is especially likely to be needed in small bandwidth situations or high dimensions. The default value is `maxk=50`.

12.3   Influence and Variance Functions

So far we have only discussed computing the fit. We are also interested in other quantities, in particular the influence and variance functions. The influence function at the fitting points is available as a direct byproduct of the fit. Since the Cholesky decomposition of the matrix $\mathbf{X}^T\mathbf{WVX}$ is available from the final iteration of the fit; $\mathbf{X}^T\mathbf{WVX} = \mathbf{R}^T\mathbf{R}$; we can rapidly compute

$$v = (\mathbf{R}^T)^{-1}e_1,$$

and hence the influence function, $\mathrm{infl}(x) = \|v\|^2 W(0)$. The variance (4.19) is slightly more work, since we must also compute the matrix

$$\mathbf{J}_2 = \mathbf{X}^T\mathbf{W}^2\mathbf{VX}.$$

To compute the fitted degrees of freedom, these quantities must be interpolated to the data points. To enable cubic interpolation, derivatives must be obtained.

Suppose x is a fitting point. For small δ, the fit at $x + \delta$ is

$$\hat{\mu}(x + \delta) = \hat{\mu}(x) + \delta\hat{\mu}'(x) + O(\delta^2)$$

and

$$\mathrm{var}(\hat{\mu}(x + \delta)) = \mathrm{var}(\hat{\mu}(x)) + 2\delta\mathrm{cov}(\hat{\mu}(x), \hat{\mu}'(x)) + O(\delta^2).$$

The derivative is

$$\frac{d}{dx}\mathrm{var}(\hat{\mu}(x)) = 2\mathrm{cov}(\hat{\mu}(x), \hat{\mu}'(x)). \tag{12.5}$$

If the local slopes are used as the derivative estimates, the covariance terms are an immediate consequence of the variance computation.

The influence function can be handled similarly:

$$\mathrm{infl}(x + \delta) \approx \mathrm{infl}(x) + \delta e_1^T(\mathbf{X}^T\mathbf{WX})^{-1}e_2 W(0). \tag{12.6}$$

Given the derivatives (12.5) and (12.6), the influence and variance functions can now be interpolated using the methods of Section 12.2.2.

12.4 Density Estimation

The local likelihood density estimate of Chapter 5 presents some additional computational difficulties. As with local likelihood regression, the local likelihood equations (5.5) can be solved numerically using the Newton-Raphson method. But the integrals must also be evaluated numerically, and this can become expensive in multiple dimensions.

If the Gaussian weight function is used, the integrals in the local likelihood equations, and the estimate itself, have closed form expressions. See Exercise 5.2 and Hjort and Jones (1996). The disadvantage of the Gaussian weight function is that the parameter space is bounded: The integral in (5.3) is infinite, if the quadratic coefficients are too large. The result is that local quadratic fitting has trouble modeling deep troughs in the density.

For local log-linear fitting with a spherically symmetric weight function, the multivariate integrals can be reduced to one dimensional ones (Exercise 12.5). These integrals are usually relatively cheap to evaluate using Taylor series.

For local log-quadratic fitting and compact weight functions, there are no general simplifications of the integrals, and the evaluation is performed using quadrature rules. An alternative is to use the product model, which replaces the spherically symmetric weights (2.11) by

$$w_i(x) = \prod_{i=1}^{d} W\left(\frac{x_{i,j} - x_{.,j}}{hs_j}\right)$$

and drops cross-product terms from the local model. This enables the multivariate integrals to be factored into one-dimensional integrals.

LOCFIT supports these simplifications, using the `itype` argument;

```
locfit(...,itype="mult")
```

Usually, this need not be given, since by default the best method for the given parameter settings will be used. The four methods are:

prod Evaluate the integrals as a product of one dimensional integrals. This is used with the product model (`kt="prod"`). The one dimensional integrals are evaluated using series expansions (for degree 0, 1 or 2 fitting) and Simpson's rule for cubic fitting.

mlin Use series expansions for multivariate local constant and log-linear models.

mult Multivariate fitting using Simpson's rule; when nothing else works.

hazd Integration method for hazard rate estimation.

Example 12.2. We generate 100 points from a trivariate standard normal model and fit local log-quadratic models. First, we fit the product model, using the compact tricube weight function. Second, we fit the full model, using the Gaussian kernel. Finally, we fit the full model using the spherically symmetric tricube kernel; this uses numerical integration. The timings, in S-4 on a 400 MHz. Pentium PC, are:

```
> x <- matrix(rnorm(300),ncol=3)
> unix.time(locfit.raw(x,kt="prod",maxk=200))
[1] 0.69 0.00 0.73 0.00 0.00
> unix.time(locfit.raw(x,kern="gauss",maxk=200))
[1] 1.08 0.02 1.10 0.00 0.00
> unix.time(locfit.raw(x,maxk=200))
[1] 35.50  0.00 35.52  0.00  0.00
```

The first fit uses the product tricube weight function and takes 0.69 seconds. The second fit uses the Gaussian weight function and takes 1.08 seconds. The third model uses the spherical tricube weight function and uses numerical integration. This takes 35.5 seconds. These times are system dependent, but the ratios should be similar on other systems.

12.5   Exercises

12.1 Recall the local likelihood $\mathcal{L}_x(a)$ defined by (4.2).

a) For any vectors a and v, and scalar λ, show

$$\frac{\partial}{\partial \lambda} \mathcal{L}_x(a + \lambda v)\Big|_{\lambda=0} = \left\langle v, \mathbf{X}^T \mathbf{W} \dot{l}(Y, \mathbf{X}a) \right\rangle.$$

b) For $v = \mathbf{J}^{-1} \mathbf{X}^T \mathbf{W} \dot{l}(Y, \mathbf{X}a)$, show this reduces to

$$(\mathbf{X}^T \mathbf{W} \dot{l}(Y, \mathbf{X}a))^T \mathbf{J}^{-1} \mathbf{X}^T \mathbf{W} \dot{l}(Y, \mathbf{X}a).$$

Assuming $\mathbf{J}$ is positive definite, show the vector $\nabla a^{(k)}$ produced by the Newton-Raphson algorithm is an ascent direction.

12.2 Consider local constant regression (with the Gaussian likelihood) for censored data. Suppose $\sigma = 1$ is known.

a) Using the updating scheme based on (7.9) and (7.11) locally at each fitting point, show that the iterations for $\hat{\mu}(x)$ satisfy

$$\hat{\mu}^{(j+1)}(x) = \hat{\mu}^{(j)}(x) + \frac{1}{\sum_{i=1}^{n} w_i(x)} \sum_{i=1}^{n} w_i(x) u_i \qquad (12.7)$$

where

$$u_i = \begin{cases} \frac{\phi(Y_i - \hat{\mu}^{(j)}(x))}{1 - \Phi(Y_i - \hat{\mu}^{(j)}(x))} & Y_i = c_i \\ (Y_i - \hat{\mu}^{(j)}(x)) & Y_i < c_i \end{cases}.$$

b) Using the Newton-Raphson algorithm, show the iterations for $\hat{\mu}(x)$ are

$$\hat{\mu}^{(j+1)}(x) = \hat{\mu}^{(j)}(x) + \frac{1}{D} \sum_{i=1}^{n} w_i(x) u_i$$

where

$$D = -\sum_{i=1}^{n} w_i(x) \ddot{l}(Y_i, c_i, \hat{\mu}^{(j)}(x)).$$

c) Show $0 \le -\ddot{l}(y, c, \mu) \le 1$ for all y, c, μ (the algebra is difficult; you may want to use plots at some steps). Hence, show the step sizes taken by (12.7) are too small, compared to the Newton-Raphson algorithm.

12.3 Consider one dimensional interpolation on an interval $[v_0, v_1]$, based on values $\hat{\mu}(v_0)$, $\hat{\mu}(v_1)$, $\hat{\mu}'(v_0)$ and $\hat{\mu}'(v_1)$.

a) The values $\hat{\mu}(v_0)$, $\hat{\mu}(v_1)$ and $\hat{\mu}'(v_0)$ define a unique quadratic polynomial $\hat{\mu}_0(x)$ for $v_0 \le v \le v_1$. Derive an expression for this quadratic polynomial. Similarly, derive the quadratic interpolant $\hat{\mu}_1(x)$ based on $\hat{\mu}(v_0)$, $\hat{\mu}(v_1)$ and $\hat{\mu}'(v_1)$.

b) Show the cubic interpolant (12.4) can be expressed as

$$\hat{\mu}(x) = (1 - \lambda)\hat{\mu}_0(x) + \lambda \hat{\mu}_1(x)$$

where $\lambda = (x - v_0)/(v_1 - v_0)$ (this shouldn't require much algebra; just verify the boundary conditions).

12.4 The object of this exercise is to compare the accuracy of direct fitting and the interpolated fits for local regression models. Use the loss measure

$$d(\hat{\mu}, \hat{\mu}^*) = \sum_{i=1}^{n} |\hat{\mu}(x_i) - \hat{\mu}^*(x_i)|$$

where $\hat{\mu}$ is the direct fit (ev="data") and $\hat{\mu}^*$ is the interpolated fit.

a) For the ethanol dataset, compute the fit of Example 3.1, using ev="data". Also compute the fit and discrepancy $d(\hat{\mu}, \hat{\mu}^*)$, using ev="tree" and a range of values of the cut parameter. Plot $d(\hat{\mu}, \hat{\mu}^*)$ against the number of fit points (in LOCFIT, this can be obtained as summary(fit)@nv). Repeat for other evaluation structures, such as ev="grid" and ev="kdtree", and compare the results.

b) Repeat for fits from other examples in this book, such as the bivariate Example 3.7 or the adaptive fit in Example 11.4.

12.5 Consider local log-linear density estimation in d dimensions, with a spherically symmetric kernel. We need to evaluate the integral $\int W(\|x\|/h)A(x)A(x)^T e^{a+b^T x} dx$, with the integrals taken over $\mathcal{R}^d$.

a) Using a canonical rotation and symmetry, express this integral in terms of the four integrals

$$\int \begin{pmatrix} 1 \\ x_1 \\ x_1^2 \\ x_2^2 \end{pmatrix} W(\|x\|/h) e^{a+\|b\|x_1} dx.$$

b) Expanding the exponentials in a Taylor series, reduce these integrals to sums of positive terms, with terms like $\int x_1^{2j} W(x/h) dx$ and $\int x_1^{2j} x_2^2 W(x/h) dx$.

c) Let

$$I_d(j_1, \ldots, j_d) = \int_{S_d} |x_1|^{j_1} \ldots |x_d|^{j_d} dS_d$$

where S_d is the surface of the unit sphere in R^d. Show that

$$\int |x_1|^{j_1} \ldots |x_d|^{j_d} W\left(\frac{\|x\|}{h}\right) dx_1 \ldots dx_d$$

$$= I_d(j_1, \ldots, j_d) \int_0^\infty r^{j_1+\ldots+j_d+d-1} W\left(\frac{r}{h}\right) dr.$$

d) Show that

$$I_d(j_1, \ldots, j_d)$$
$$= \frac{\Gamma(\frac{j_d+1}{2})\Gamma(\frac{j_1+\ldots+j_{d-1}+d-1}{2})}{\Gamma(\frac{j_1+\ldots+j_d+d}{2})} I_{d-1}(j_1, \ldots, j_{d-1})$$
$$= 2\frac{\Gamma(\frac{j_1+1}{2})\ldots\Gamma(\frac{j_d+1}{2})}{\Gamma(\frac{j_1+\ldots+j_d+d}{2})}.$$

The beta integral, $\int_0^1 x^{a-1}(1-x)^{b-1} dx = \Gamma(a)\Gamma(b)/\Gamma(a+b)$, may be assumed.

13

Optimizing Local Regression

Previous chapters have developed local regression and likelihood in many settings and provided considerable intuitive motivation and examples to support the methods. In this chapter, more formal results characterizing the performance of local regression are reviewed and developed.

Section 13.1 studies rates of convergence for local regression and their optimality properties. Section 13.2 studies optimal constants and efficiency of the weight functions. Section 13.3 develops *finite sample* minimax properties of local regression. Section 13.4 studies design adaptation properties of the minimax estimates and discusses some important points about model indexing.

13.1 Optimal Rates of Convergence

The pioneering results for asymptotic optimality of local regression was the series of papers by Stone (1977, 1980, 1982). These papers in turn established consistency, optimality of pointwise rates of convergence and optimality of global rates of convergence, and the results in this section largely follow Stone's work. Earlier results for density estimation can be found in Farrell (1972).

When discussing optimality properties, one has to carefully consider what is required. For example, it is meaningless to talk about a single estimation problem. A perfectly legitimate estimate in the regression problem is to ignore the data and take $\hat{\mu}(x)$ to be an arbitrary function, say $\hat{\mu}(x) =$

$e^{-2x} + x^3$. This estimate will beat any data-based estimate, if the true mean function happens to be $\mu(x) = e^{-2x} + x^3$. On the other hand, this estimate will be poor if $\mu(x)$ happens to be anything else.

Thus a local regression estimate cannot be optimal in any sense, if we consider the performance one function at a time. Rather, the simultaneous performance of the estimate over a class of candidates for $\mu(x)$ must be considered, for example, by requiring uniformity over the class. In this chapter we consider classes of the form

$$\mathcal{F}_{M,p,x} = \left\{ \mu : |\mu(y) - \langle g, A(y - x) \rangle| \leq \frac{M^{p+1}}{(p+1)!} \|y - x\|^{p+1} \forall y \right\} \quad (13.1)$$

where g is the vector of Taylor series coefficients of $\mu(x)$ up to order p. The class $\mathcal{F}_{M,p,x}$ is a superset of all $(p+1)$ times differentiable functions whose $(p+1)$st derivative is bounded by M.

A sequence of estimates $\hat{\mu}_n(x)$ is said to have rate of convergence $n^{-\alpha}$; $\hat{\mu}_n(x) - \mu(x) = O(n^{-\alpha})$ in probability, if

$$\lim_{c \to \infty} \limsup_{n \to \infty} \sup_{\mu \in \mathcal{F}_{M,p}} P(|\hat{\mu}_n(x) - \mu(x)| > cn^{-\alpha}) = 0. \quad (13.2)$$

If $n^\alpha(\hat{\mu}_n(x) - \mu(x))$ has a limiting distribution,

$$n^\alpha(\hat{\mu}_n(x) - \mu(x)) \Rightarrow Z,$$

then $\lim_{n \to \infty} P(|\hat{\mu}_n(x) - \mu(x)| > cn^{-\alpha}) = P(|Z| > c)$, and clearly (13.2) holds. However, the definition (13.2) is weaker; roughly, it says the distribution of $n^\alpha(\hat{\mu}_n(x) - \mu(x))$ must not drift off to ∞.

Example 13.1. For any local regression estimate $\hat{\mu}_n(x)$, Chebychev's inequality implies that

$$P(|\hat{\mu}_n(x) - \mu(x)| > cn^{-2/5})$$
$$\leq \frac{\mathrm{E}\left((\hat{\mu}_n(x) - \mu(x))^2\right)}{c^2 n^{-4/5}}$$
$$= \frac{n^{4/5}\mathrm{var}(\hat{\mu}_n(x)) + n^{4/5}\mathrm{bias}(\hat{\mu}_n(x))^2}{c^2}. \quad (13.3)$$

Now suppose x is one dimensional, $\hat{\mu}_n(x)$ is local linear regression and the bandwidth is $h = n^{-1/5}$. By (2.44) and (2.45), both the variance and *squared* bias are of size $n^{-4/5}$. In fact, the bias is uniform over the class $\mathcal{F}_{M,1,x}$:

$$\sup_{\mu \in \mathcal{F}_{M,1,x}} |\mathrm{E}(\hat{\mu}_n(x)) - \mu(x)| \leq \frac{M}{2} \sum_{i=1}^{n} |l_i(x)|(x_i - x)^2 = O(n^{-2/5}).$$

This implies that the right-hand side of (13.3) has a finite limit, which tends to 0 as $c \to \infty$. Thus, $\hat{\mu}_n(x) - \mu(x) = O(n^{-2/5})$ in probability.

Generalizing this example to arbitrary dimension d and degree p leads to the following result.

Theorem 13.1 Suppose $\mu(x)$ has bounded partial derivatives of order $p+1$ and the estimate $\hat{\mu}_n(x)$ is local regression of degree p with bandwidth $h_n = n^{-1/(2p+2+d)}$. Fix x with $f(x)$ bounded away from 0 in a neighborhood of x. Under the basic design assumption (2.36),

$$\hat{\mu}_n(x) - \mu(x) = O(n^{-\alpha})$$

in probability, where

$$\alpha = \frac{p+1}{2p+2+d}. \tag{13.4}$$

Theorem 13.1 derives the rate of convergence achieved by a local regression estimate. It remains to show that this is the best possible rate of convergence, as provided by the following definition.

Definition 13.1 The rate $n^{-\alpha}$ is

- **achievable** if there exists a sequence of estimates $\hat{\mu}_n(x)$ for which $\hat{\mu}_n(x) - \mu(x) = O(n^{-\alpha})$ in probability $\mathcal{F}_{M,p,x}$.

- a **lower bound** on the rate of convergence if no sequence of estimates converges faster than $n^{-\alpha}$ uniformly over $\mathcal{F}_{M,p,x}$. That is, no sequence of estimates satisfies

$$\limsup_{n \to \infty} \sup_{\mu \in \mathcal{F}_{M,p,x}} P(|\hat{\mu}_n(x) - \mu(x)| > cn^{-\alpha}) = 0$$

for all $c > 0$.

- the **optimal rate of convergence** if it is both achievable and a lower bound.

By Theorem 13.1, local regression of degree p achieves the rate of convergence $n^{-\alpha}$ with $\alpha = (p+1)/(2p+2+d)$ over $\mathcal{F}_{M,p,x}$. The following theorem shows that this rate is also a lower bound and therefore optimal.

Theorem 13.2 Consider the regression model (2.1) and suppose the errors ϵ_i are independent with the $N(0,1)$ distribution. Let $h_n = n^{-1/(2p+2+d)}$, and suppose the design sequence satisfies the basic assumption (2.36) with $h = h_n$. Then $n^{-\alpha}$, with α given by (13.4), is the optimal rate of convergence over the class $\mathcal{F}_{M,p}$.

Proof: For simplicity, suppose we are fitting at $x = 0$. The proof is for $d = 1$ and $p = 1$; the general case is left as an exercise.

Let $\mu_0(x) = 0$ for all x and choose $\mu_1 \in \mathcal{F}_{M,1,0}$ with $\mu_1(u) = 0$ for $|u| > 1$ and $\mu_1(0) > 0$. A satisfactory choice is

$$\mu_1(x) = \begin{cases} \frac{M}{\pi^2}(1 + \cos(\pi x)) & |x| \leq 1 \\ 0 & |x| > 1 \end{cases}.$$

For a sequence h_n, define the scaled $\mu_{1,n}(x) = h_n^2 \mu_1(x/h_n)$. Note that $\mu_{1,n} \in \mathcal{F}_{M,p}$ for all n.

Consider the problem of testing the hypotheses

$$\mathcal{H}_0 \quad : \quad \mu = \mu_0$$
$$\text{vs. } \mathcal{H}_1 \quad : \quad \mu = \mu_{1,n}. \tag{13.5}$$

Given a sequence of estimates $\hat{\mu}_n(x)$, one possible test statistic is $\hat{\mu}_n(0)$, with the decision rule

$$\delta(x) = \begin{cases} 0 & \hat{\mu}_n(0) \leq \mu_{1,n}(0)/2 \\ 1 & \hat{\mu}_n(0) > \mu_{1,n}(0)/2 \end{cases}. \tag{13.6}$$

Clearly, if $\hat{\mu}_n(0)$ has convergence $o_p(h_n^2)$, then this test will be consistent; that is, both error probabilities tend to 0.

Conversely, consider the sequence of likelihood ratio tests for these hypotheses. By the Neyman-Pearson lemma (Lehmann, 1986, Section 3.2) these tests are most powerful. If the likelihood ratio tests cannot consistently distinguish between the hypotheses (13.5), then no other test - in particular, the test (13.6) - can consistently distinguish the hypotheses. In this case, h_n^2 is a lower bound on the rate of convergence.

The log-likelihood ratio test for (13.5) rejects $\mathcal{H}_0$ for large values Λ_n, where

$$\begin{aligned} 2\log(\Lambda_n) &= \sum_{i=1}^{n} Y_i^2 - \sum_{i=1}^{n}(Y_i - \mu_{1,n}(x_i))^2 \\ &= 2\sum_{i=1}^{n}\mu_{1,n}(x_i)Y_i - \sum_{i=1}^{n}\mu_{1,n}(x_i)^2. \end{aligned}$$

Let $T_n = \sum_{i=1}^{n}\mu_{1,n}(x_i)^2$. Then $2\log(\Lambda_n)$ is normally distributed, with mean $-T_n$ under $\mathcal{H}_0$ and T_n under $\mathcal{H}_1$. The variance is

$$\text{var}\,(2\log(\Lambda_n)) = 4\sum_{i=1}^{n}\mu_{1,n}(x_i)^2 = 4T_n.$$

Clearly, for the test to consistently distinguish between the hypotheses requires the standard deviation $2\sqrt{T_n}$ to be of smaller order of magnitude than T_n; that is, $T_n \to \infty$. But

$$\frac{T_n}{nh_n^5} = \frac{1}{nh_n}\sum_{i=1}^{n}\mu_1\left(\frac{x_i}{h_n}\right)^2 \to f(0)\int \mu_1(x)^2 dx.$$

In particular, if $h_n = n^{-1/5}$, T_n remains finite and the hypotheses cannot be consistently distinguished. Consequently, $h_n^2 = n^{-2/5}$ is a lower bound on the rate of convergence of $\hat{\mu}(x)$. □

Stone (1982) also develops a theory for optimal global rates of convergence. This considers L_q-norm loss functions:

$$\left(\int |\hat{\mu}(x) - \mu(x)|^q \right)^{1/q}, \tag{13.7}$$

where the integral is taken over a compact set, typically the domain of the data. The special case $q = 2$ corresponds to mean integrated squared error. For $1 < q < \infty$, the optimal rates match the pointwise rates (Theorem 13.1) under some regularity conditions. As $q \to \infty$, the loss (13.7) is essentially the maximal deviation $\sup_x |\hat{\mu}(x) - \mu(x)|$. Under this loss, Stone shows the optimal rate, achieved by local regression, is

$$\sup_x |\hat{\mu}(x) - \mu(x)| = O_p \left(\left(\frac{\log(n)}{n} \right)^\alpha \right), \tag{13.8}$$

which is slightly slower than the rates in Theorem 13.1. Konakov and Piterbarg (1984) and Härdle and Luckhaus (1984) also discuss uniform convergence of smooth regression estimates. The corresponding result for density estimation is in Khas'minskii (1978).

Uniform convergence results such as (13.8) are relevant to several results in this book. First, it provides lower bounds on the size of simultaneous confidence bands (Section 9.2.2). Second, the classification error probabilities in Section 8.5.2 require uniform convergence of the local likelihood estimate. A more subtle application was in Exercise 6.5, where the consistency of the change point estimates requires $\hat{\Delta}_t \to 0$ uniformly away from the change point.

Other types of convergence results can also be considered. Katkovnik (1983), Ioffe and Katkovnik (1989), Devroye, Györfi, Krzyzak and Lugosi (1994) and others consider strong convergence, or convergence with probability 1, in various settings. Horváth (1991) and references therein provide results on the asymptotic distribution of the L_q-norm for density estimation.

13.2   Optimal Constants

Consider one dimensional local linear regression with unbounded support. The asymptotic bias is

$$\mathrm{E}(\hat{\mu}(x)) - \mu(x) = \frac{1}{2} h^2 \mu''(x) \frac{\int v^2 W(v) dv}{\int W(v) dv} + o(h^2) \tag{13.9}$$

and the asymptotic variance is

$$\text{var}(\hat{\mu}(x)) = \frac{\sigma^2}{nhf(x)} \frac{\int W(v)^2}{(\int W(v)dv)^2} + o((nh)^{-1}). \tag{13.10}$$

Suppose the weight function satisfies

$$\int W(v)dv = 1$$

$$\int v^2 W(v)dv = 1 \tag{13.11}$$

(for non-negative weight functions, this can always be achieved by rescaling). Let $W_0(v) = c(1 - v^2/k^2)_+$ with $k^2 = 15/4$ and $c = 3/(4k)$. Then $W_0(v)$ is the version of the quadratic weight function rescaled to satisfy (13.11).

Let $W(v)$ be any other non-negative weight function satisfying (13.11); this implies that the asymptotic biases (13.9) for $W(v)$ and $W_0(v)$ are equal. We have

$$\int W(v)W_0(v)dv \geq c \int W(v)(1 - x^2/k^2)dv = c(1 - 1/k^2)$$

with equality when $W = W_0$. In particular, this implies $\int W_0(v)(W(v) - W_0(v))dv \geq 0$. Thus,

$$\begin{aligned}
\int W(v)^2 dv &= \int (W_0(v) + W(v) - W_0(v))^2 dv \\
&= \int W_0(v)^2 dv + \int (W_0(v) - W(v))^2 dv \\
&\quad +2 \int W_0(v)(W(v) - W_0(v))dv \\
&\geq \int W_0(v)^2 dv.
\end{aligned}$$

Thus, the quadratic weight function, appropriately scaled, is optimal in the sense that any weight function producing the same asymptotic bias has larger asymptotic variance. This result was originally due to Epanechnikov (1969).

How good are other weight functions? From (13.9) and (13.10), the asymptotic mean squared error is

$$\text{ASE}(h) = \frac{h^4 \mu''(x)^2}{4} \frac{(\int v^2 W(v)dv)^2}{(\int W(v)dv)^2} + \frac{\sigma^2}{nhf(x)} \frac{\int W(v)^2 dv}{(\int W(v)dv)^2}.$$

Minimizing over h, the 'optimal' bandwidth is

$$h^5 = \frac{\sigma^2 \int W(v)^2 dv}{nf(x)\mu''(x)^2(\int v^2 W(v)dv)^2} \tag{13.12}$$

	$\text{eff}_2(W)$	$\text{eff}_4(W)$
Epanechnikov	1.000	1.000
Tricube	0.998	0.997
Bisquare	0.994	0.994
Triweight	0.987	0.987
Triangle	0.986	0.990
Gaussian	0.951	0.932
Rectangular	0.930	0.959

TABLE 13.1. Asymptotic pointwise efficiencies of weight functions for local linear and local cubic smoothing.

and the corresponding mean squared error is

$$\min_h \text{ASE}(h) = \frac{5\mu''(x)^{2/5}\sigma^{8/5}}{4n^{4/5}f(x)^{4/5}} \frac{(\int W(v)^2 dv)^{4/5}(\int v^2 W(v)dv)^{2/5}}{(\int W(v)dv)^2}.$$

For Epanechnikov's weight function, the component dependent on W is

$$\frac{(\int W(v)^2 dv)^{4/5}(\int v^2 W(v)dv)^{2/5}}{(\int W(v)dv)^2} = \left(\frac{6}{\sqrt{125}}\right)^{4/5}.$$

The efficiency of a weight function is defined as the ratio of sample sizes to obtain the same asymptotic MSE as Epanechnikov's weight function:

$$\text{eff}_2(W) = \frac{6\left(\int W(v)dv\right)^{5/2}}{\sqrt{125}\left(\int v^2 W(v)dv\right)^{1/2}\int W(v)^2 dv}.$$

Similar calculations can be performed for local cubic fitting. The minimized asymptotic mean squared error is

$$\frac{9\mu^{(4)}(x)^{2/9}\sigma^{16/9}}{8 \cdot 72^{1/9}(nf(x))^{8/9}}\left(\int W^*(v)^2 dv\right)^{8/9}\left|\int v^4 W^*(v)dv\right|^{2/9}$$

where $W^*(v) = e_1^T \mathbf{M}_1^{-1} A(v) W(v)$ is the equivalent kernel (2.40). Epanechnikov's kernel $W(v) = (1-v^2)I_{[0,1]}(v)$ is again optimal, and the asymptotic efficiency of other weight functions is

$$\text{eff}_4(W) = \frac{5}{4 \times 21^{1/4}}\left(\int W^*(v)^2 dv\right)^{-1}\left|\int v^4 W^*(v)dv\right|^{-1/4}.$$

Table 13.1 summarizes the asymptotic efficiencies of weight functions supported in LOCFIT. The bisquare, tricube and triweight weight functions are all very efficient, and their smoothness makes them preferable to the Epanechnikov weights in practice. The rectangular and Gaussian weights are slightly less efficient but are sometimes preferred for simplicity.

13.3 Minimax Local Regression

Consider a linear estimate $\hat{\mu}(x) = \sum_{i=1}^{n} l_i(x)Y_i$. Recalling the mean and variance expressions (2.13), the mean squared error of the estimate is

$$R(l(x), \mu) = \sigma^2 \sum_{i=1}^{n} l_i(x)^2 + \left(\sum_{i=1}^{n} l_i(x)\mu(x_i) - \mu(x) \right)^2.$$

What is the best possible weight diagram $l(x)$? This can't be answered directly, since it depends on the unknown $\mu(x)$. Rather, we have to consider simultaneous performance over a class of possible mean functions. In this section we find $l(x)$ to solve the minimax problem:

$$\min_{\{l(x)\}} \sup_{\mu \in \mathcal{F}_{M,p,x}} R(l(x), \mu) \tag{13.13}$$

where $\mathcal{F}_{M,p,x}$ is defined by (13.1).

Similar results in related settings are given in Legostaeva and Shiryayev (1971) and Sacks and Ylvisaker (1978, 1981). An asymptotic variant of the results for local regression was studied in Fan (1993). Specifically, he showed that the local linear smoother with Epanechnikov's weight function is the minimax linear estimate. This is much more powerful than the results of the previous section; the previous results only showed that Epanechnikov's weight function was the optimal choice for the local linear estimate. For other asymptotic minimax results in regression problems, see Brown and Low (1991), Low (1993) and references therein.

The requirement that $R(l(x), \mu)$ be bounded over $\mathcal{F}_{M,p,x}$ is equivalent to the conditions

$$\sum_{i=1}^{n} l_i(x)(x_i - x)^j = \begin{cases} 1 & j = 0 \\ 0 & 1 \le j \le p \end{cases} . \tag{13.14}$$

Henderson's theorem (Theorem 2.1) then implies the minimax smoother is a local polynomial fit of degree p, provided the condition on sign changes is satisfied.

By considering the mean function

$$\mu(x_i) = \frac{M}{(p+1)!} \|x_i - x\|^{p+1} \text{sgn}(l_i(x)),$$

we can reduce (13.13) to $\min_{\{l(x)\}} R_{\text{sup}}(l(x))$, where

$$R_{\text{sup}}(l(x)) = \sigma^2 \sum_{i=1}^{n} l_i(x)^2 + \frac{M^2}{(p+1)!^2} \left(\sum_{i=1}^{n} |l_i(x)| \|x_i - x\|^{p+1} \right)^2. \tag{13.15}$$

The minimum is now taken over all weight diagrams satisfying the constraints (13.14). The following theorem shows the solution has the form

$$
\begin{aligned}
l_i^{\alpha,\gamma}(x) &= (\langle \alpha, A(x_i - x) \rangle - \gamma \|x_i - x\|^{p+1})_+ \\
&\quad - (\langle \alpha, A(x_i - x) \rangle + \gamma \|x_i - x\|^{p+1})_-
\end{aligned}
\tag{13.16}
$$

for an appropriate vector α and constant γ.

Theorem 13.3 For any γ, choose $\alpha = \alpha(\gamma)$ so that $\{l_i^{\alpha,\gamma}(x)\}$ satisfies (13.14). Suppose also, there exists a γ such that

$$
\gamma = \frac{M^2}{\sigma^2 (p+1)!^2} \sum_{i=1}^{n} |l_i^{\alpha,\gamma}(x)| \|x_i - x\|^{p+1}.
\tag{13.17}
$$

Then $l_i^{\alpha,\gamma}(x)$ is the minimax optimal weight diagram.

Remark. Clearly $\gamma \geq 0$; this implies that for any x_i, at most one of the components on the right of (13.16) is nonzero. If $\langle \alpha, A(x_i - x) \rangle > 0$, then $l_i^{\alpha,\gamma}(x) \geq 0$. If $\langle \alpha, A(x_i - x) \rangle < 0$, then $l_i^{\alpha,\gamma}(x) \leq 0$. Thus, $\{l_i^{\alpha,\gamma}(x)\}$ has sign changes corresponding to the zeros of $\langle \alpha, A(u - x) \rangle$, and by Henderson's theorem, $\{l_i^{\alpha,\gamma}(x)\}$ is a local polynomial estimate of degree p. Explicitly, the non-negative weights are

$$
w_i(x) = \left(1 - \gamma \frac{\|x_i - x\|^{p+1}}{|\langle \alpha, A(x_i - x) \rangle|} \right)_+.
$$

Example 13.2. Suppose $x_i = i/n; i = \ldots, -1, 0, 1, \ldots$ and $x = 0$ (the doubly infinite sequence avoids boundary effects). Consider the case $p = 1$ and $\alpha = (\alpha_0, \alpha_1)$. By symmetry, $\alpha_1 = 0$, and

$$
l_i^{\alpha,\gamma}(x) = (\alpha_0 - \gamma x_i^2)_+ = \alpha_0 \left(1 - \frac{\gamma}{\alpha_0} \frac{i^2}{n^2} \right)_+.
$$

That is, the minimax weight diagram is Epanechnikov's weight function, with bandwidth $h = \sqrt{\alpha_0/\gamma}$. Since

$$
\sum_{i=1}^{n} x_i^2 l_i(x) \approx n\alpha_0 \int_{-h}^{h} x^2 (1 - (x/h)^2) dx = \frac{4}{15} n\alpha_0 h^3
$$

and $\gamma = \alpha_0 / h^2$, (13.17) is approximately

$$
\frac{\alpha_0}{h^2} = \frac{M^2}{4\sigma^2} \frac{4}{15} n\alpha_0 h^3,
$$

giving

$$
h^5 = \frac{15\sigma^2}{nM^2}.
$$

This replicates the optimal bandwidth (13.12), with $\mu''(x) = M$, $f(x) = 1$ and $W(v) = (1 - v^2)_+$.

Proof: The proof we give here is based on Lagrange multipliers, although the reader may object to this since (13.15) is not differentiable. An alternative longer proof involves substituting the claimed solution into (13.15) and showing that this is a minimum.

Consider the quantity

$$\frac{1}{2\sigma^2} R_{\text{sup}}(l(x)) - \left\langle \alpha, \sum_{i=1}^{n} l_i(x) A(x_i - x) \right\rangle ;$$

α is the vector of Lagrange multipliers. Differentiating with respect to $l_j(x)$ yields the equation

$$l_j(x) + \gamma \cdot \text{sgn}(l_j(x)) \|x_j - x\|^{p+1} - \langle \alpha, A(x_j - x) \rangle = 0$$

with γ defined by 13.17. This gives

$$l_j(x) = \begin{cases} \langle \alpha, A(x_j - x) \rangle - \gamma \|x_i - x\|^{p+1} & \text{if } +\text{ve} \\ \langle \alpha, A(x_j - x) \rangle + \gamma \|x_i - x\|^{p+1} & \text{if } -\text{ve} \end{cases}$$

with α chosen to satisfy the constraints (13.14).   □

13.3.1 Implementation

The minimax weight diagrams are found numerically. The implementation has two steps:

1. For any γ, find $\alpha(\gamma)$ so that $\{l_i^{\alpha,\gamma}(x)\}$ satisfies (13.14).

2. Find the γ satisfying (13.17).

The constraints (13.14) can be written in the form

$$e_1 + \gamma \sum_{l_i^{\alpha,\gamma}(x) \neq 0} \text{sgn}(l_i^{\alpha,\gamma}(x)) |x_i - x|^{p+1} = \sum_{l_i^{\alpha,\gamma}(x) \neq 0} A(x_i - x) A(x_i - x)^T \alpha.$$

This is not a linear system of equations, since the domain of the sums depend on α. But it provides the basis for an iterative algorithm: From an initial α, evaluate the sums, solve for α, and iterate to convergence. In fact, this is the Newton-Raphson algorithm for minimizing the convex function

$$\mathcal{L}(\alpha) = \frac{1}{2} \sum_{i=1}^{n} l_i^{\alpha,\gamma}(x)^2 - \alpha_0.$$

The minimax weights are implemented in LOCFIT, using kern="minmax". Of course, M/σ is usually unknown and needs to be specified. But this can be treated as a smoothing parameter, and cross validation and other diagnostics can help choose this.

The minimax weights are useful in small samples with nonuniform designs. In these cases, both constant and nearest neighbor bandwidths often have problems with sparse neighborhoods, and produce unsatisfactory fits. Locally adaptive bandwidths, such as those considered in Chapter 11, will also be unsatisfactory, since there is insufficient data for the local criteria to work well.

Example 13.3. The mmsamp datset is generated with $n = 50$, $x_i \sim U[0,1]$ and

$$Y_i = 2 - 5x_i + 5\exp(-(20x_i - 10)^2) + \epsilon_i$$

with $\epsilon_i \sim N(0,1)$. The mean function is essentially linear, with a sharp peak at $x = 0.5$. This model was considered by Seifert and Gasser (1996) with smaller error variance.

The maximum (absolute) second derivative is $M = 4000$. The minimax fit is computed by setting kern="minmax":

```
> fit1 <- locfit(y~x, data=mmsamp, deg=1, kern="minmax",
+    alpha=4000, ev="grid", mg=100, flim=c(0,1))
```

With mimimax weights, alpha is interpreted as M/σ, rather than as a nearest neighbor bandwidth. For comparison, the fit with constant bandwidth $h = 0.05$ is also computed:

```
> fit2 <- locfit(y~x, data=mmsamp, deg=1, alpha=c(0,0.05),
+    ev="grid", mg=100, flim=c(0,1))
```

The bandwidth was chosen to match the fitted degrees of freedom; both fits produce $\nu_2 = 19$.

The fits in Figure 13.1 are clearly a mess. This isn't surprising, since extremely small bandwidths are being used to model the peak. But the fit with constant h is quite unreasonable, with several spurious sharp peaks. The problem is that in some regions, the constant bandwidth is essentially a linear *extrapolant* from nearby points. The design adaptation of the minimax weights nicely solves the problem.

Remark. One might try to smooth the mmsamp dataset with the locally adaptive bandwidth rules of Chapter 11, since with knowledge of the true mean, the bandwidths should clearly be much larger near the boundaries. But this is unlikely to be successful: If the adaptive rule is sensitive enough to detect the true peak (represented by a single observation), then other clusters of observations may be falsely detected.

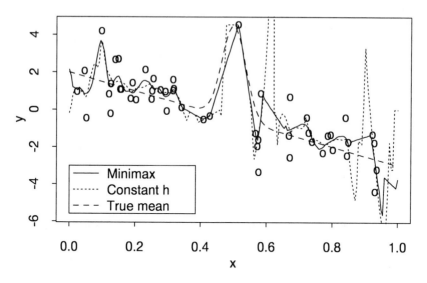

FIGURE 13.1. Smoothing a dataset with sparse data regions and small bandwidths.

13.4 Design Adaptation and Model Indexing

The success of the minimax methods in the previous section centers around the use of the smoothness class $\mathcal{F}_{M,p,x}$ and then allowing the minimax optimality criterion to dictate the choice of smoothing weights. The ratio M/σ plays the role of the smoothing parameter: Changing M/σ controls the bias-variance trade-off and the roughness of the resulting estimate. The ratio can be interpreted as a signal-to-noise ratio. M represents a bound on the signal, as measured by the $(p+1)$st derivative, and σ represents the noise.

The minimax method leads to the rather obscure form of the smoothing weights. The computation can be tricky when M/σ is large and the resulting fit should be close to interpolation. But this is precisely the case for which design adaptation provided by the minimax method should be most useful.

An alternative is to solve a restricted version of the minimax problem, where (13.15) is minimized over a smaller class of weight diagrams. For example, consider only the weight diagrams resulting from standard local regression weighting schemes, and choose the resulting variable bandwidth $h = h(x)$ to minimize the risk function $R_{\sup}(l(x))$. We refer to the resulting estimate as M-indexed local regression. Like nearest neighbor and constant bandwidth methods, the M-indexed method is linear: The smoothing weights depend on the design points x_i and the bandwidth h, but not on the responses Y_i.

The M-indexed estimate can be implemented along the lines of the locally adaptive bandwidth algorithm in Section 11.2: Begin with a small bandwidth and increase until the minimum risk is found. This is implemented in LOCFIT by setting the adaptive criterion `acri="mindex"`. The *third* component of the smoothing parameter `alpha` is then interpreted as M/σ.

To illustrate the performance of this method, we present a simulation study of the model from Example 13.3.

Example 13.4. 1000 datasets are generated from the model used in example 13.3 (using a new set of design points x_i for each dataset). For each dataset, local linear smooths are computed by the minimax, M-indexed and constant h methods. The smooths are evaluated at 101 points, equally spaced on $[0, 1]$, and the sum of squared errors is computed. The sum of squared errors is then averaged over the 1000 replications. Thus, the final loss is

$$\frac{1}{1000} \sum_{i=1}^{1000} \sum_{j=0}^{100} (\hat{\mu}(v_j) - \mu(v_j))^2$$

with $v_j = j/100$. The loss is computed for various values of the smoothing parameters M and h.

Figure 13.2 shows the results. The correspondence $h^5 = 1.36608/M^2$ is used, motivated by (2.46) with $\mu''(x)$ replaced by M. The minimax and M-indexed methods produce almost identical results, indicating that there is very little loss in using M-indexing in place of the minimax method. The results with constant h are vastly inferior. Note also that the best mean squared errors are obtained at $M = 500$, whereas $\sup |\mu''(x)| = 4000$.

It should be noted that both the M-indexed and constant h curves in Figure 13.2 are computed with the same estimates: Local linear estimates with the tricube weight function. What changes is how the estimates are averaged. For the M-indexed method, estimates are averaged with the *same* assumption about the true mean function.

By contrast, under the constant h method, fits are being averaged with completely different assumptions and amounts of smoothing: An interval $(x - h(x), x + h(x))$ may contain no observations for some realizations and several observations for other realizations. The results indicate the severe problem of the constant bandwidth specification rather than problems with local regression as Seifert and Gasser (1996) attempted to argue.

Gu (1998) has extensively discussed model averaging, and in particular emphasized the point that averaging must be with respect to an *assumption about the smoothness of the underlying function* rather than the smoothing parameter. But Gu specified smoothness assumptions in the form $\int f''(x)^2 < \lambda$ and obtained smoothing splines as the optimal solution.

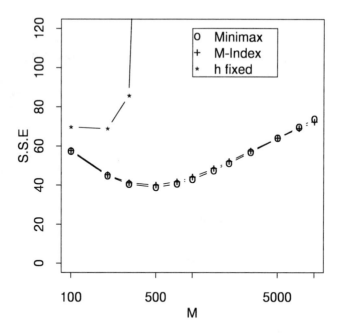

FIGURE 13.2. Average sum of squared errors for constant bandwidths, minimax bandwidths and minimax local regression.

The importance of using finite sample variances and risk, rather than asymptotic approximations, must also be emphasized. The asymptotic minimax arguments (Fan 1993 and Example 13.2) yield, for a uniform design, a constant bandwidth estimator. But for the problem in Example 13.4, the asymptotic theory is clearly inadequate.

13.5 Exercises

13.1 Suppose we have observations $(x_i, Y_i); i = 1, \ldots, n$ with the sequence $\{x_i\}$ and bandwidth h satisfying the basic assumption (2.36). Suppose also $E(Y_i|x_i) = \mu(x_i)$ and $var(Y_i|x_i) = \sigma^2(x_i)$; both $\mu(\cdot)$ and $\sigma^2(\cdot)$ are continuous functions. Let $S_n = \sum_{i=1}^{n} K(\frac{x_i - x}{h}) Y_i$. Show that

$$E(S_n/nh) = \mu(x)f(x) \int K(v) dv + o(1)$$

$$var(S_n/nh) = \frac{f(x)\sigma^2(x)}{nh} \int K(v)^2 dv + o((nh)^{-1})$$

and hence $S_n/(nh)$ converges in probability to $f(v)\mu(x) \int K(v) dv$.

13.2 Complete the proof of Theorem 13.2 for general p and d.

13.3 Consider local regression at a point x. Suppose the design density satisfies $f(u) \sim (u - x)^\alpha$ as $u \to x$, with α fixed.

   a) Under the condition $nh_n^{\alpha+1} \to \infty$, provide a modification of the basic assumption (2.36).

   b) Derive first order asymptotic approximations to the bias and variance; show these are of size $O(h_n^{p+1})$ and $O(1/(nh_n^{\alpha+1}))$ respectively. Hence derive the optimal size of h_n and obtain the best possible rate of convergence achieved by the local polynomial estimate.

   c) Show this rate is optimal for all estimates.

13.4 (Double smoothing, continued)

   a) For the double smooth in Exercise 2.5, the weight diagram is, for large h, approximately $W^*(i/h)$ where

$$W^*(x) = \begin{cases} \frac{1}{2} + \frac{|x|}{4} & |x| \leq 1 \\ -\frac{1}{2} + \frac{|x|}{4} & 1 < |x| < 2 \\ 0 & \text{elsewhere} \end{cases} .$$

   Show $W^*(x)$ is a fourth order kernel ($\int v^2 W^*(v) = 0$), and the asymptotic efficiency of this weight function is 0.775.

   b) Some authors recommend different bandwidths h_1 and h_2 at the two stages of the double smooth. Show the weight diagram for this double smooth is approximately

$$W^*(x) = W_{h_1}(x) + W_{h_2}(x) - (W_{h_1} * W_{h_2})(x).$$

   Assuming $W(x)$ is a symmetric second order kernel, show $W^*(x)$ is a fourth order kernel.

   c) Taking $h_1 = 1$, $h_2 = h$ and $W(u) = \exp(-u^2/2)/\sqrt{2\pi}$, evaluate the asymptotic efficiency as a function of h. Show that the maximum efficiency, 0.918, is attained at $h = 1$ and the efficiency tends to 0 as $h \to 0$ or $h \to \infty$.

   d) Using a symbolic algebra package such as Maple, perform similar computations for the compactly supported weight functions.

13.5 For the mmsamp dataset used in Examples 13.3 and 13.4, compute and plot the minimax and M-indexed local linear fits for values of M ranging from 100 to 4000. Plot the fits, and compute the sum of squared errors over an equally spaced grid. Repeat for constant and nearest neighbor bandwidths over a comparable range of smoothing parameters, and compare the results.

Appendix A
Installing LOCFIT in R, S and S-Plus

The LOCFIT S interface was developed in version 4 of S (Chambers 1998). The software is also compatible with versions of S-Plus that support, as a minimum, the level of S functionality described in Becker, Chambers and Wilks (1988) and Chambers and Hastie (1992). LOCFIT has been successfully used with S-Plus versions 3.3, 3.4, 4.0 and 5.0. Users of S-Plus 4.5 will need the professional version; the standard version lacks the standard S programming interface, and LOCFIT will be difficult or impossible to use in this environment.

The free R package (Ihaka and Gentleman 1996) is also largely compatible with (although developed independently of) S version 3. LOCFIT has been successfully compiled and run with R version 0.63.

This appendix describes installation for various environments: S-Plus 3.3 and 4.0 for Windows (Section A.1); S-Plus 3.3 and 3.4 for UNIX (Section A.2); S-Plus 5.0 for UNIX (Section A.3) and R 0.63 (Section A.4).

The LOCFIT software can be obtained from the World-Wide Web at the address

> http://cm.bell-labs.com/stat/project/locfit

A.1 Installation, S-Plus for Windows

1. Download the self-extracting zip file for S-Plus 3.3 or S-Plus 4.0 as appropriate. Save to your Temp directory.

2. Determine your SHOME folder. The default is C:\spluswin for S-Plus 3.3 and C:\program files\splus4 for S-Plus 4.0.

3. Run the self-extracting zip file. Extract to SHOME\library\locfit.

4. (Optional) Add the line

   ```
   locfit          Local Regression and likelihood
   ```

   to the file SHOME\library\readme.txt.

LOCFIT is now ready to run. Start an S-Plus session, and type

```
> library("locfit",first=T)
```

to attach the LOCFIT library.

A.2 Installation, S-Plus 3, UNIX

Installation in S-Plus 3.3 and 3.4 requires compiling the C code and using dyn.load.shared(), dyn.load() or dyn.load2(), depending on what is supported on your system. As distributed, the installation files are set up to make a shared library, for use with dyn.load.shared().

1. Save the locfit.shar file in a new locfit directory (if you are a site administrator, use $SHOME/library/locfit). Unpack the file with

   ```
   % sh locfit.shar
   ```

2. If your SHOME is *not* /usr/splus (type Splus SHOME to be sure), you need to change SHOME at the top of Makefile.S3.

3. Install with the command

   ```
   % sh install.S3
   ```

4. On some systems, the shared library file may be called locfit.sl, rather than locfit.so. In this case, you'll need to change LOCFIT's .First.lib() function. In S-Plus:

   ```
   > .First.lib() <- function(library, section)
   +    dyn.load.shared(paste(library, section,
   +       "locfit.sl", sep = "/"))
   ```

5. (For site administrators) Add the line

   ```
   locfit          Local Regression and Likelihood
   ```

to the file /usr/splus/library/README.

6. LOCFIT should now be ready to use. Start S-Plus (in another directory) and attach the library with

   ```
   > library("locfit",first=T,lib.loc=getenv("HOME"))
   ```

   Note the lib.loc argument should be the parent of your locfit
   directory; this can be omitted if you followed the site administrator
   installation.

7. (Optional; check LOCFIT is working first.) Remove the source files.
   On your LOCFIT installation directory, remove everything **except** the
   locfit.so file and the .Data subdirectory and contents.

For systems that don't support shared libraries, but support dyn.load()
or dyn.load2(), edit Makefile.S3 and change the WHICH_LOAD lines to

```
# WHICH_LOAD=static.load
WHICH_LOAD=dyn.load
# WHICH_LOAD=dyn.load.shared
```

Then run the install.S3 script. This will build the object file locfit_1.o.
Run S-Plus, and change the .First.lib() function to

```
> .First.lib() <- function(library, section)
+ dyn.load(paste(library, section, "locfit_1.o", sep = "/"))
```

A.3 Installation, S-Plus 5.0

S-Plus 5.0 is based on S version 4; some advanced features of LOCFIT require
these versions of S and S-Plus to operate correctly. At the time of writing,
the routines for installing libraries in S-Plus 5.0 are unreliable; see the
LOCFIT web page for any updates.

1. Unpack the LOCFIT source archive in a clean directory. If possible,
   use $SHOME/library/locfit.

2. Ensure the SHOME environment variable is properly set.

3. Run the install.S5 script:

   ```
   % sh install.S5
   ```

   This should make some subdirectories, and compile the necessary
   source files.

4. Start S-Plus, and source the S files:

```
> source("locfit.s")
> source("locfitdat.s")
> source("locfit4.s")
> rm(.First.lib)
```

LOCFIT should now be ready to use. From another directory, the library is attached with

```
> library("locfit",first=T)
```

with a lib.loc argument if necessary.

A.4   Installing in R

R is a free statistical programming language, available from the Comprehensive R Archive Network (CRAN) archive,

http://lib.stat.cmu.edu/R/CRAN

The R langauge is very similar to S version 3, and most features of LOCFIT will work with R. The installation instructions here are designed for UNIX systems. For Windows users, compiled versions of the LOCFIT library are usually available from CRAN.

To install the LOCFIT library in R 0.63:

1. Download the locfitR.tar.gz file. Save in a convenient place, such as $RHOME/src/library.

2. Unpack and install with:

```
% gunzip locfitR.tar.gz
% tar -xvf locfitR.tar
% R INSTALL locfit
```

To access LOCFIT from R:

```
% R
> library("locfit")
> data(ethanol)
> fit <- locfit(NOx~E,data=ethanol)
> plot(fit,get.data=T)
```

LOCFIT in R is largely compatible with S version 3. The major exception (as of R version 0.63) is the lack of a Trellis library for displaying fits in three of more dimensions. Datasets must be explicitly loaded before use; data(ethanol) in the above example.

Appendix B
Additional Features: LOCFIT in S

This appendix describes a number of features of LOCFIT in S that haven't been described elsewhere in this book. Section B.1 covers the `predict()` and `preplot()` functions. Section B.2.2 discusses iterative procedures using LOCFIT. Section B.3 covers arithmetic operators for `"locfit"` objects and other uses of S version 4 classes and methods. Section B.4 discusses some ways to interface between LOCFIT and trellis displays.

B.1 Prediction

As described in Chapter 12, LOCFIT selects a small number of points at which to compute the local fit. Two functions, `predict.locfit()` and `preplot.locfit()`, are used to interpolate the fit to additional points. The difference:

- `predict.locfit()` returns a vector of the interpolated values.

- `preplot.locfit()` returns an object with the `"preplot.locfit"` class, which contains the interpolated values and some additional information.

As its name suggests, `preplot.locfit()` is a preliminary step for plotting a `"locfit"` object; `plot.locfit()` simply sets up a grid of prediction points, calls `preplot.locfit()` and plots the resulting `"preplot.locfit"` object. In fact, `plot(fit)` is largely equivalent to `plot(preplot(fit))`.

The arguments to `preplot.locfit()` and `predict.locfit()` are similar. The two most important arguments are the `object` and `newdata`. The `newdata` can be specified in several forms. A vector of points may be provided for a one dimensional fit or matrix of points for a multidimensional fit. Alternatively, a data frame can be provided; the names must match the variables used in the original fit. A list can also be provided as `newdata`: In this case, prediction is performed on a grid, with the components of the list representing the grid margins. If no `newdata` is provided, prediction is at the fit points.

An alternative to `newdata` is `where`, a character string controlling generation of a set of prediction points. Allowed values include `where="grid"` (a grid generated by the `lfmarg()` function), `where="data"` (the data points) and `where="fitp"` (the selected fitting points).

Other arguments for `predict.locfit()` and `preplot.locfit()` include

- `what=` controls what is predicted: `"coef"` (the function values, default), `"infl"` (the influence function), `"nlx"` ($\|l(x)\|$), `"band"` (the bandwidth) and `"deg"` (the local polynomial degree).

- `band=` Computes standard errors and confidence bands for the fit. Available choices are `"none"` (the default), `"global"`, `"local"` and `"prediction"`.

- `tr` a back-transformation function. The default is the inverse of the link function, so the prediction will be for the mean function for regression models or the density for density estimation.

- `get.data=` If `TRUE`, the data will be stored on the result. This is most useful for plotting, since the data points are then added to the plot.

B.2   Calling `locfit()`

There are many instances where one needs to write functions that call `locfit()` and do some additional processing before returning required results. Examples include `gcv()` and other model assessment functions in Section 3.4.2. We may also want to write functions that iterate through several calls to `locfit()`, for example, the `locfit.robust()` function mentioned in Section 6.4. This section discusses implementation of such functions.

B.2.1   Extracting from a Fit

A first attempt to implement the `gcv()` function may be

```
gcv <- function(...)
{ fit <- locfit(...)
```

```
    n <- fit@mi["n"]
    df <- fit@dp["df1"]
    -2 * n * fit@dp["lk"]/(n-df)^2
}
```

This will work, but problems arise with the approach. The L_1 cross validation function (Exercise 3.3) needs a call to `residuals()`; modifying the preceding code to this case will not work properly.

A more robust approach is to construct a call to the `locfit()` function. The precise implementation of `gcv()` is

```
gcv <- function(x,...)
{
    m <- match.call()
    if(is.numeric(x))
        m[[1]] <- as.name("locfit.raw")
    else {
        m[[1]] <- as.name("locfit")
        names(m)[2] <- "formula"
    }
    fit <- eval(m, sys.parent())
    dp <- fit@dp
    z <- dp[c("lk", "df1", "df2")]
    n <- fit@mi["n"]
    z <- c(z, (-2 * n * z[1])/(n - z[2])^2)
    names(z) <- c("lik", "infl", "vari", "aic")
    z
}
```

The `match.call()` line simply returns a list containing all the arguments provided to the `gcv()` function. This is converted into either a `locfit()` or `locfit.raw()` call, depending on the type of the first argument. The fit is evaluated by

```
eval(m, sys.parent())
```

and the returned object is the standard `locfit()` object from which the desired information can be extracted.

B.2.2 Iterative Use of `locfit()`

In a number of applications, we want to call `locfit()` repeatedly to arrive at a single fit, changing some of the variables (for example, the response or prior weights) at each iteration. One such example is the LOWESS style robustness iterations described in Section 6.4, where prior weights must be determined at each iteration based on residuals from the previous fit. Another example is the censored regression model from Section 7.2, where the response must be updated at each iteration, estimating the mean residual life from the previous fit. A third example is the quasi-likelihood procedure (Section 4.3.4) for observations with unequal variances.

These problems can be solved by performing the iterations locally for each fitting point. LOCFIT also provides functions to do the iterations globally: `locfit.robust()` for robust regression, `locfit.censor()` for censored regression and `locfit.quasi()` for quasi-likelihood. This approach allows more control over the procedure: for example, to change the weighting scheme for robust regression or to change the mean residual life estimate for censored regression.

How are these functions written? One could write a loop to call either `locfit()` or `locfit.raw()`; the second is easier and avoids repeated interpretation of the model formula (an expensive operation). Let's look at the `locfit.quasi()` function:

```
> locfit.quasi <-
function(x, y, weights, ..., iter = 3,
  var = function(mean) abs(mean))
{
  m <- match.call()
  n <- length(y)
  w0 <- lfq.wt <- if(missing(weights)) rep(1, n) else weights
  m[[1]] <- as.name("locfit.raw")
  for(i in 0:iter) {
    m$weights <- lfq.wt
    fit <- eval(m, sys.parent())
    fh <- fitted(fit)
    lfq.wt <- w0/var(fh)
  }
  fit
}
```

This function accepts the same arguments as `locfit.raw()`. At each iteration, it computes the weights `lfq.wt` and substitutes these in the call to `locfit.raw()`. There are two additional arguments: `iter` to set the number of iterations and `var` to specify the variance function (expect trouble, if this can return negative numbers!).

Remark. The `locfit.quasi()` function actually has additional code not shown above, so it can be called with a model formula. The implementation in this case is to recall `locfit()`, setting the `lfproc` argument to `locfit.quasi()`.

B.3   Arithmetic Operators and Math Functions

In version 4 S, arithmetic operators such as +, comparison operators such as < and mathematical functions such as log() are all generic functions: Methods can be defined so that these functions can operate on non-numeric data. For a complete description of these capabilities, see Chapter 8 of Chambers (1998).

The "Math" group of functions contains log() and other common functions. A method is provided for the "preplot.locfit" class:

```
> setMethod("Math", signature(x="preplot.locfit"), lfm)
```

where the method lfm() is defined as

```
> lfm <- function(x)
{ fit = x@trans(x@fit)
  x@fit = callGeneric(as.numeric(fit))
  x@trans = function(x) x
  x
}
```

Thus,

```
> fit1 <- locfit(NOx~E, data=ethanol, alpha=0.5)
> pred <- log(preplot(fit1, lfmarg(fit1)))
> plot(pred)
```

will plot the log of the fitted curve.

Suppose we write a new math function, expit() to compute the function $e^x/(1+e^x)$. This must be written in a stable form to avoid unnecessary overflow:

```
> expit <- function(x)
{ y <- numeric(length(x))
  y[x>0] <- 1/(1+exp(-x[x>0]))
  z <- exp(x[x<=0])
  y[x<=0] <- z/(1+z)
  y
}
```

As written, this function will only work for numeric data. To ensure it works for "preplot.locfit" objects (and any other classes for which "Math" methods are defined), use the command

```
> setGeneric("expit",group="Math")
```

The use of "Math" functions has a couple of deficiencies that should be noted. First, no attempt is made to preserve standard errors for the transformed values. Second, if the data is stored on the "preplot.locfit" object, this will not be transformed.

Methods are also defined for the operator group `"Ops"`, which includes both arithmetic and comparison operators. These differ from `"Math"` functions in that two operands are required and either one or both of these may be `"preplot.locfit"` objects. For example, suppose we want to average two different fits:

```
> fit1 <- locfit(NOx~E,data=ethanol,alpha=0.5)
> fit2 <- locfit(NOx~E,data=ethanol,alpha=0.7)
> newx <- lfmarg(fit1)
> pred1 <- preplot(fit1,newx)
> pred2 <- preplot(fit2,newx)
> plot((pred1+pred2)/2)
```

Here, we compute fits on the ethanol dataset with two different bandwidths, make predictions on a grid of points and average the two predictions.

A limited number of arithmetic operators are also available to operate directly on a `"locfit"` object, as summarized in Table B.1. For likelihood models, note these methods are implemented directly on the estimation scale, unlike the operators for `"preplot.locfit"` objects.

e1	e2	Operators
`"locfit"`	`"locfit"`	$+,-$
`"locfit"`	`"numeric"`	$+,-,{}^*,/$
`"numeric"`	`"locfit"`	$+,-,{}^*$

TABLE B.1. Operators available for `"locfit"` objects.

B.4   Trellis Tricks

In several examples, we have used trellis displays, where each panel contains a smooth of the same dataset, but with different smoothing parameters. Directly producing these using `xyplot()` is difficult, since one would have to replicate the original data frame an appropriate number of times, then add additional level variables.

A shortcut is to set up dummy variables to set up the Trellis display, and get the real data inside the panel function. Thus, the real code producing Figure 2.3 is

```
> a <- c(0.2,0.4,0.6,0.8)
> anames <- paste("a =",a)
> xyplot(a~I(a)|as.factor(anames)),
+    xlim=range(ethanol$E), ylim=range(ethanol$NOx),
+    panel = function(x, ...)
+    { xd <- ethanol$E
```

```
+      yd <- ethanol$NOx
+      fit <- locfit.raw(xd, yd, alpha=x)
+      panel.xyplot(xd, yd, cex = 0.7)
+      lines(fit)
+   },
+   strip=strip.loc, ylab="NOx",
+   xlab = "Equivalence Ratio")
```

The vector a contains the four smoothing parameters and is used in the
model formula. Thus, when Trellis calls the panel function, the x argument
will be the smoothing parameter. Thus, the panel function must first get
the real data (xd and yd) and call locfit.raw() using alpha=x as the
smoothing parameter.

Appendix C
C-LOCFIT

The examples in this book have included the S code for implementation with LOCFIT. It is also possible to run LOCFIT as a stand-alone program, known as as C-LOCFIT.

C.1 Installation

C.1.1 Windows 95, 98 and NT

You should have the self-extracting zip file. Run this file (if it didn't run automatically when downloading); the contents should be extracted to a C:\locfit folder. The locfit executable is C:\locfit\locfit.exe; the example datasets should be automatically installed in the LFData subfolder.

C.1.2 UNIX

You should have the LOCFIT source archive. Unpack this file in a new LOCFIT directory. Before compiling, the correct version must be defined in the local.h file. Edit this file, so the first few lines are

```
#define CVERSION
/* #define RVERSION */
/* #define SVERSION */
/* #define INTERFACE */
```

Then at the UNIX prompt, type

```
% make locfit
```

Remark. To compile C-LOCFIT *without* code for the X Window System, edit the `Makefile` and replace the line

```
COBJ= main.o cmd.o lfd.o pout.o arith.o
```

by

```
COBJ= mainnox.o cmd.o lfd.o pout.o arith.o
```

Also, set `LDFLAGS=-lm`. The default graphics will then be ASCII style.

Assuming compilation was successful, install the example datasets:

```
% locfit
locfit> run install.cmd
locfit> exit
```

To make C-LOCFIT accessible to multiple users, move the `locfit.sh` script to `/usr/local/bin/locfit`. You'll need to edit the `LFHOME` variable in that script.

C.2   Using C-LOCFIT

To start C-LOCFIT, type

```
% locfit
```

at the UNIX prompt (Windows users should click the LOCFIT icon in the installation directory). You should obtain the prompt

```
locfit>
```

A large number of commands are available; some are described in the remainder of this appendix. The easiest way to explore the capabilities is the `example` command:

```
locfit> example
Example 3.1. Local Regression with plot
Example 3.3. Local Regression; print results
Example 3.4. Residual plots for Local Regression
...
```

With no arguments, `example` just prints one-line descriptions of the available examples. The numbers match the example numbers in this book. To obtain the code for an example, simply give the example number:

```
locfit> example 3.1

Example 3.1. Local Regression with plot
```

```
locfit NOx~E data=ethanol alpha=0.5
plotfit data=T
```

The arguments to C-LOCFIT's locfit function are in direct correspondence to the S version arguments. To actually run the example, type

```
locfit> example 3.1 run
```

C.2.1   Data in C-LOCFIT

In its simplest use, C-LOCFIT is a calculator:

```
locfit> 3*4+6
18.00000
```

Results can also be assigned to variables:

```
locfit> c1=3*4+6
locfit> c1
18.00000
```

Creating and assigning vectors is easy:

```
locfit> x=seq(0,1,5) y=10*sqrt(x)+rnorm(5)
locfit> x y
 0.00000   -0.06061
 0.25000    6.43409
 0.50000    7.75524
 0.75000   10.47189
 1.00000   10.37475
```

This creates five equally spaced points on $[0,1]$ for the independent (x) variable and a response vector as $y_i = 10\sqrt{x_i} + \epsilon_i$ with ϵ_i standard normal noise.

The readfile command reads data from a file:

```
locfit> ethanol=readfile file=ethanol.dat NOx C E
```

This reads the file ethanol.dat (the file= can be omitted, as long as the filename is given as the first argument), creating three variables: NOx, C and E. The dataset is saved (in a binary format) as a C-LOCFIT dataset, in the file LFData/ethanol.lfd. One could also achieve this in two separate commands:

```
locfit> readfile file=ethanol.dat NOx C E
locfit> ethanol=savedata NOx C E
```

Another optional argument to readfile is arith. If arith=T, then each item in the input file is parsed by C-LOCFIT's arithmetic interpreter before being assigned to the data matrix. If the input file is numeric, this has little effect other than to slow things down slightly. One use of the `arith` argument is processing categorical input:

```
locfit> versicolor=0 virginica=1
locfit> iris=readfile iris.dat arith=T spec sl sw pl pw
```

Here, the input file contains a categorical variable, `spec`, that can't be handled by C-LOCFIT. By first assigning numeric values to the two species and then specifying `arith=T`, the `species` variable will be created with 0-1 values.

In addition to user-defined variables, a number of predefined constants point to components of the fit: `alpha`, `h`, `pen` (the three components of the smoothing parameter), `like` (the likelihood at the fitted values), `infl`, `vari` (fitted degrees of freedom by the $\mathrm{tr}(\mathbf{L})$ and $\mathrm{tr}(\mathbf{L}^T\mathbf{L})$ definitions) and `resv` (estimate of the residual variance). These constants can also be assigned to, although should only be done with extreme care. More usefully, these constants can be used in computations:

```
locfit> locfit NOx~E data=ethanol alpha=0.7
locfit> alpha vari 88*(-2*like)/(88-infl)^2
      0.7  4.76494 0.151757
```

Here, a local quadratic model is fitted to the ethanol dataset. Then, the smoothing parameter, fitted degrees of freedom and the generalized cross validation statistic are printed.

Recognized components of arithmetic expressions include:

- Floating point numbers.

- Parenthesized expressions such as `4*(3+5)`.

- Arithmetic operators `+,-,*,/` and `^`. Precedence of `*,/` over `+,-` is recognized; the `^` (exponentation) operator has top precedence.

- Comparisons `<,>,<=,>=,==` and `<>`.

- The `,` operator for catenation. Lists of numbers in a function argument must be parenthesized. For example, `min(1,2)` will interpret 1 and 2 as two separate arguments and produce an error message. The correct call is `min((1,2))`.

- Subsets. For example, `x[3,5,2]` will return the third, fifth and second elements of `x`.

- Common mathematical functions such as `sin()` and `sqrt()`, and summary functions including `min()`, `max()`, `sum()` and `mean()`.

- Random number generators: `rexp()`, `rnorm()`, `runif()`, `rpois()` for exponential, normal, uniform and Poisson random variables. All expect the sample size as the first argument; `rpois` is given the mean as the second argument.

- sample(x,n) draws random sample of size n (with replacement) from x.

C.3   Fitting with C-LOCFIT

The main purpose of C-LOCFIT is local fitting, using the locfit command. The simplest use is

```
locfit> readdata ethanol
locfit> locfit NOx~E
```

where the first argument NOx~E specifies a model formula. The model formula is similar to that in the S version: NOx~E+C specifies a model with two predictors; ~E specifies a density estimation model with no response. One difference is that variables cannot be created in the formula. Thus log(NOx)~E is not valid; instead, use

```
locfit> readdata ethanol
locfit> NOx=log(NOx)
locfit> locfit logNOx~E
```

The relocfit command is used to recompute a fit with new parameters:

```
locfit> readdata ethanol
locfit> locfit NOx~E alpha=0.6 deg=1
locfit> plotfit data=T
locfit> # try smaller value of alpha
locfit> relocfit alpha=0.3
locfit> plotfit data=T
```

The locfit command has numerous optional arguments, mostly in one-to-one correspondence with the same arguments in the S version. See the online examples for usage.

Fits can be saved in a binary format. The commands are savefit and readfit, to save and read the fit respectively. For example,

```
locfit> readdata ethanol
locfit> locfit NOx~E alpha=0.5
locfit> savefit ethanol
```

creates a file LFPlot/ethanol.fit. savefit can also be accomplished by assigning the locfit command:

```
locfit> ethanol=locfit NOx~E alpha=0.5 data=ethanol
```

C.4 Prediction

C-LOCFIT includes `predict`, `fitted` and `residuals` commands that are similar to the S version methods. Thus

```
locfit> readdata ethanol
locfit> locfit NOx~E alpha=0.5
Evaluation structure 1, 8 points.
locfit> predict E=0.6,0.8,1.0
  0.600000       0.723941
  0.800000       2.754441
  1.000000       3.118365
```

interpolates the fit to the three points 0.6, 0.8 and 1.0. Alternatively,

```
locfit> predict where=data
  0.907000       3.745299
  0.761000       2.244242
  1.108000       1.418052
...
```

gives fitted values at the data points. There is also the `what` argument that is similar to the S version (Section B.1). For example, `what=nlx` interpolates $\|l(x)\|$. The `fitted` and `residuals` commands produce fitted values and residuals respectively; these have optional `type` and `cv` arguments.

C.5 Some additional commands

The following commands can also be used:

`outf filename a` Sets output file for results, default is standard output. Important output is sent to this file, while working output, such as the prompt and error messages, are still be sent to terminal. If the `a` is specified, output will be appended to the file. The default is to overwrite.

`run file.cmd` Read commands from the command file `file.cmd`.

`for i=1:10` A for loop, repeats following commands ten times. The loop is terminated by a `endfor` command. Any numeric variable can be used as the loop variable; `for x=rnorm(100)` is legitimate, and `x` loops through 100 standard normal variables.

`seed -(ThcjK2` Set the seed for the random number generator. A string of eight characters should be used.

`exit` exits LOCFIT.

Appendix D
Plots from C-LOCFIT

C-LOCFIT plots are produced using the `plotdata` and `plotfit` commands. By default, plots are produced in a C-LOCFIT plot window. If the `fmt=post` argument is given, a PostScript file, `lfplot.ps`, is generated:

```
locfit> locfit NOx~E data=ethanol
locfit> # plot the fit on the screen.
locfit> plotfit
locfit> # generate a PostScript file
locfit> plotfit fmt=post
```

Two more advanced plot commands are `setplot` and `track`. `setplot` is used to automatically generate a plot when a model is fitted. For example,

```
locfit> setplot 0 plotfit
```

will automatically plot the fit using the `plotfit` command every time a fit is generated (this command is included in the default `LFInit.cmd` file, which is executed at startup). `track` is used to accumulate a set of points - for example, a goodness of fit statistic - over successive calls of the `locfit` command.

Under the X Window System, a plot will be damaged when you move or resize graphics windows. Pressing `return` should result in a redraw. If it doesn't, force a redraw with the `replot` command:

```
locfit> replot win=0
```

`replot` can also change some graphical parameters, including plot styles and the viewpoint for three dimensional plots.

D.1 The `plotdata` Command

The `plotdata` command produces scatter plots of the data:

```
locfit> readdata ethanol
locfit> plotdata E NOx
locfit> plotdata x=E y=C z=NOx
```

In C-LOCFIT, all plots are set up with three axes (x, y and z). Two dimensional plots are simply produced by setting the viewing angle so that one axis is not visible.

The first `plotdata` command above produces an $x - y$ scatter plot of the E and NOx variables. The second `plotdata` produces a three dimensional scatterplot. The view point can be set with the argument `view=a1,a2` with longitude `a1` and latitude `a2`. `view=0,0` views from the North pole, displaying the x-y plane. The default (when `x`, `y` and `z` arguments are all provided) is `view=45,45`.

An optional `type=` argument controls the style of the plot:

```
locfit> readdata ethanol
locfit> plotdata E NOx type=p
locfit> plotdata E NOx type=l
locfit> plotdata E NOx type=b
locfit> plotdata E NOx C type=q view=0,0
```

The first three types, `p`, `l` and `b` stand for points, lines and both, respectively. `type=q` color codes points according to a greyscale, colored according to levels of the z variable (`C` in this example).

A C-LOCFIT plot can be built of several components, each with its own x, y and z variables. To add a component to an existing plot, use `add=T`:

```
locfit> readdata ethanol
locfit> plotdata E NOx
locfit> locfit NOx~E
locfit> fit=fitted
locfit> plotdata E fit add=T type=s
```

This first plots the data. Then fitted values from a local regression fit are computed and added to the plot. `type=s` results in segments joining the new component (the fit) with the previous component (the data).

D.2 The `plotfit` Command

The `plotfit` command is used to plot fits. Its arguments largely correspond to the S version `plot.locfit()` function.

Example D.1. A local quadratic fit is computed and plotted:

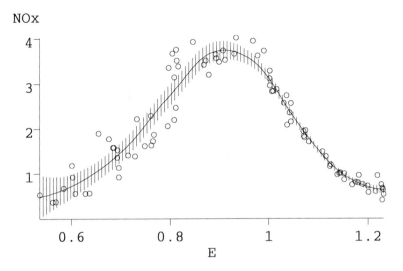

FIGURE D.1. Plotting a fit, with data and error bars.

```
locfit> locfit NOx~E data=ethanol alpha=0.5
locfit> plotfit band=1 data=T type=lnsp fmt=post w=112 h=75
```

This requests that both confidence intervals (band=1) and data (data=T) be added to the plot. Thus the plot consists of four components: the fit, the lower confidence limits, the upper confidence limits and the data. type=lnsp specifies plot styles for the four components: lines, none, segments (joining upper and lower limits) and points. Figure D.1 shows the result. Note that fmt=post w=112 h=75 specifies a PostScript file with dimensions 112 × 75 millemetres.

Important: Fits are plotted in the x-z plane, rather than the x-y plane. Thus, adding data to the plot separately:

```
locfit> plotfit
locfit> plotdata E NOx add=T
```

will not produce the desired result. Instead, use

```
locfit> plotfit
locfit> plotdata E z=NOx add=T
```

For fits with two predictor variables, the default plot type is a contour plot. Plot types supported for two dimensional predictors are

```
locfit> plotfit type=c        Contour Plot
locfit> plotfit type=w        Wire Frame
locfit> plotfit type=i        Colored level plot
```

Density

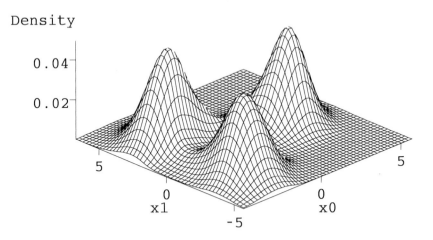

FIGURE D.2. C-LOCFIT plot of a two dimensional density

Example D.2. The trimodal dataset from Figure 5.4 is fitted with a bandwidth 0.35:

```
locfit> locfit ~x0+x1 alpha=0.35 data=trimod
locfit> plotfit type=w
```

The wireframe plot is shown in Figure D.2.

Confidence bands can be added to the two predictor plots, but the result will usually be messy. The most satisfactory results may be obtained with type=wns on a low resolution plot:

```
locfit> plotfit band=g type=wns m=20
```

Generally, confidence intervals are better viewed through one dimensional cross sections.

Fit surfaces with three or more predictors can't be directly plotted, but one or two dimensional cross sections can be. For example,

```
locfit> x1=runif(100) x2=runif(100) x3=runif(100)
locfit> locfit ~x1+x2+x3
locfit> plotfit x2=0
```

produces a contour plot of the fit in x1 and x3, with x2 fixed at 0.

D.3   Other Plot Options

A number of other options for controlling features such as labels and colors are available. These can be given as arguments to `plotfit`, `plotdata` or `replot` commands:

- `win=3`. C-LOCFIT supports up to five graphics windows, numbered 0, 1, 2, 3 and 4. The `win=` argument controls which window is used.

- `add=T` causes the plot to be added to an existing plot.

- `split=3,4,5` For contour plots, the `split` argument is used to specify contour levels.

- `xlim`, `ylim` and `zlim` specify ranges for the axis. Note, when using `plotfit` these do not affect the domain over which the plot is computed. Thus

  ```
  locfit> locfit NOx~E data=ethanol
  locfit> plotfit E=0.75,1
  ```

  would be more desirable than

  ```
  locfit> plotfit xlim=0.75,1
  ```

- `xlab`, `ylab`, `zlab` and `main` give labels for the x, y and z axis, and a main title. Note spaces cannot be used in these arguments, but underscores (_) will be converted to spaces.

Colors. Colors available in LOCFIT are white, black, red, green, blue, magenta, yellow and cyan. There is also a greyscale used in image plots; the colors used in the greyscale can be set using

```
locfit> greyscale red blue
```

to set the greyscale running from red to blue (ideally, this should be set in the `LFInit.cmd` file). To change colors used for other components of the plot, use the `setcolor` command:

```
locfit> setcolor back=black fore=red text=green
```

will set the background color to black, the foreground color to red, and the text color to green. Available arguments to `setcolor` are `back` (background), `fore` (everything else except `patch2`), `axis`, `text`, `lines`, `points`, `cont` (contours), `clab` (contour labels), `seg` (segments), `patch1` and `patch2` (used in wireframe plots).

References

Aitchison, J. and C. G. C. Aitken (1976). Multivariate binary discrimination by the kernel method. *Biometrika* **63**, 413–420.

Akaike, H. (1973). Information theory and an extension of the maximum likelihood principle. In B. N. Petrov and F. Csàki (Eds.), *Second International Symposium on Information Theory*, Budapest, pp. 267–281. Akademia Kiadó.

Akaike, H. (1974). A new look at the statistical model identification. *IEEE Transactions on Automatic Control* **19**, 716–723.

Aldous, D. (1989). *Probability Approximations via the Poisson Clumping Heuristic*. New York: Springer-Verlag.

Alekseev, V. G. (1982). Nonparametric estimates of probability density and its derivatives. Проблемы Передачи Информации *(Problems of Information Transmission)* **18**(2), 22–29 (100–106).

Allen, D. M. (1974). The relationship between variable selection and data augmentation and a method of prediction. *Technometrics* **16**, 125–127.

Anderson, J. A. (1972). Separate sample logistic discrimination. *Biometrika* **59**, 19–35.

Anderson, T. W. (1971). *The Statistical Analysis of Time Series*. New York: Wiley.

Andrews, D. F. and A. M. Herzberg (1985). *Data*. New York: Springer-Verlag.

Antoniadis, A., G. Grégoire and G. Nason (1999). Density and hazard rate estimation for right censored data using wavelet methods. *Journal of the Royal Statistical Society, Series B* **61**, 63–84.

Atkeson, C. G., A. W. Moore and S. Schaal (1997). Locally weighted learning. *AI Review* **11**, 11–73.

Azzalini, A. and A. W. Bowman (1993). On the use of nonparametric regression for checking linear relationships. *Journal of the Royal Statistical Society, Series B* **55**, 549–557.

Azzalini, A., A. W. Bowman and W. Härdle (1989). On the use of nonparametric regression for model checking. *Biometrika* **76**, 1–11.

Bartlett, M. S. (1950). Periodogram analysis and continuous spectra. *Biometrika* **37**, 1–16.

Becker, R. A., J. M. Chambers and A. R. Wilks (1988). *The New S Language: A Programming Environment for Data Analysis and Graphics*. Pacific Grove, California: Wadsworth & Brooks-Cole.

Becker, R. A., W. S. Cleveland, M. J. Shyu and S. P. Kaluzny (1994). Trellis displays: User's guide. Technical report, AT&T Bell Laboratories.

Bickel, P. J. and M. Rosenblatt (1973). On some global measures of the deviations of density function estimates. *The Annals of Statistics* **1**, 1071–1095.

Blackman, R. B. and J. W. Tukey (1958). *The Measurement of Power Spectra from the Point of View of Communication Engineering*. New York: Dover.

Boden, T. A., R. J. Sepanski and F. W. Stoss (1992). Trends '91: A compendium of data on global change - highlights. Technical report, Carbon Dioxide Information Analysis Center, Oak Ridge National Laboratory, Oak Ridge, Tennessee.

Borgan, Ø. (1979). On the theory of moving average graduation. *Scandinavian Actuarial Journal* **1979**, 83–105.

Bowman, A. W. (1984). An alternative method of cross-validation for the smoothing of density estimates. *Biometrika* **71**, 353–360.

Bowman, A. W. and A. Azzalini (1997). *Applied Smoothing Techniques for Data Analysis: the Kernel Approach with S-Plus Illustrations*. Oxford University Press.

Box, G. E. P. and D. R. Cox (1964). An analysis of transformations. *Journal of the Royal Statistical Society, Series B* **26**, 211–252.

Breiman, L. and J. H. Friedman (1985). Estimating optimal transformations for multiple regression and correlation (with discussion). *Journal of the American Statistical Association* **80**, 580–619.

Breiman, L., J. H. Friedman, R. A. Olshen and C. J. Stone (1984). *Classification and Regression Trees.* Monterey: Wadsworth and Brooks/Cole.

Breiman, L. and W. S. Meisel (1976). General estimates of the intrinsic variability of data in nonlinear regression models. *Journal of the American Statistical Association* **71**, 301–307.

Brillinger, D. R. (1977). Discussion of Stone (1977). *The Annals of Statistics* **5**, 622–623.

Brinkman, N. D. (1981). Ethanol fuel - a single-cylinder engine study of efficiency and exhaust emissions. *SAE Transactions* **90**, 1410–1424.

Brown, L. D. and M. G. Low (1991). Information inequality bounds on the minimax risk (with an application to nonparametric regression). *The Annals of Statistics* **19**, 329–337.

Brown, L. D., M. G. Low and L. H. Zhao (1997). Superefficiency in nonparametric function estimation. *The Annals of Statistics* **25**, 2607–2625.

Buckley, J. and I. James (1979). Linear regression with censored data. *Biometrika* **66**, 429–436.

Buckley, M. J. and G. K. Eagleson (1989). A graphical method for estimating the residual variance in nonparametric regression. *Biometrika* **76**, 203–210.

Buja, A., T. Hastie and R. Tibshirani (1989). Linear smoothers and additive models (with discussion). *The Annals of Statistics* **17**, 453–555.

Carroll, R. J. (1982). Adapting for heteroscedasticity in linear models. *The Annals of Statistics* **10**, 1224–1233.

Chambers, J. M. (1998). *Programming with Data. A Guide to the S Language.* New York: Springer.

Chambers, J. M. and T. J. Hastie (1992). *Statistical Models in S.* Pacific Grove, California: Wadsworth & Brooks-Cole.

Cheng, B. and D. M. Titterington (1994). Neural networks: A review from a statistical perspective (with discussion). *Statistical Science* **9**, 2–54.

Chiu, S. T. (1991). Bandwidth selection for kernel density estimation. *The Annals of Statistics* **19**, 1883–1905.

Cleveland, W. S. (1979). Robust locally weighted regression and smoothing scatterplots. *Journal of the American Statistical Association* **74**, 829–836.

Cleveland, W. S. (1993). *Visualizing Data*. Summit, New Jersey: Hobart Press.

Cleveland, W. S. (1994). Coplots, nonparametric regression, and conditionally parametric fits. In T. W. Anderson, K. T. Fang and I. Olkin (Eds.), *Multivariate Analysis and its Applications*, pp. 21–36. Hayward: Institute of Mathematical Statistics.

Cleveland, W. S. and S. J. Devlin (1988). Locally weighted regression: An approach to regression analysis by local fitting. *Journal of the American Statistical Association* **83**, 596–610.

Cleveland, W. S. and E. H. Grosse (1991). Computational methods for local regression. *Statistics and Computing* **1**, 47–62.

Cleveland, W. S., E. H. Grosse and W. M. Shyu (1992). Local regression models. In J. M. Chambers and T. J. Hastie (Eds.), *Statistical Models in S*, pp. 309–376. Pacific Grove: Wadsworth and Brooks/Cole.

Cleveland, W. S. and C. R. Loader (1996). Smoothing by local regression: Principles and methods. In W. Härdle and M. G. Schimek (Eds.), *Statistical Theory and Computational Aspects of Smoothing*, Heidelberg, pp. 10–49. Physica-Verlag.

Cleveland, W. S., C. L. Mallows and J. E. McRae (1993). ATS methods: Nonparametric regression for nongaussian data. *Journal of the American Statistical Association* **88**, 821–835.

Clough, R. W. and J. L. Tocher (1965). Finite element stiffness matrices for analysis of plates in bending. In *Proc. Conf. Matrix Methods in Structural Mechanics*.

Conte, S. D. and C. de Boor (1980). *Elementary Numerical Analysis. An Algorithmic Approach* (Third ed.). McGraw-Hill.

Cook, R. D. (1977). Detection of influential observations in linear regression. *Technometrics* **19**, 15–18.

Cook, R. D. and S. Weisberg (1994). *An Introduction to Regression Graphics*. New York: John Wiley & Sons.

Cover, T. M. (1968). Rates of convergence for nearest neighbor procedures. In *International Conference on Systems Sciences*, Honolulu, pp. 413–415. Western Periodicals.

Cover, T. M. and P. E. Hart (1967). Nearest neighbor pattern classification. *IEEE Transactions on Information Theory* **IT-13**, 21–27.

Cowden, D. J. (1962). Weights for fitting polynomial secular trends. Technical Paper 4, School of Business Administration, University of North Carolina.

Cox, D. D., E. Koh, G. Wahba and B. S. Yandell (1988). Testing the (parametric) null model hypothesis in (semiparametric) partial and generalized spline models. *The Annals of Statistics* **16**, 113–119.

Cox, D. R. (1972). Regression models and life tables (with discussion). *Journal of the Royal Statistical Society, Series B* **74**, 187–220.

Cox, D. R. and P. A. W. Lewis (1966). *The Statistical Analysis of Series of Events*. London: Methuen.

Cox, D. R. and D. Oakes (1984). *Analysis of Survival Data*. London: Chapman and Hall.

Craven, P. and G. Wahba (1979). Smoothing noisy data with spline functions. *Numerische Mathematik* **31**, 377–403.

Daniell, P. J. (1946). Discussion on Symposium on Autocorrelation in Time Series. *Supplement to Journal of the Royal Statistical Society* **8**, 88–90.

Daniels, H. E. (1962). The estimation of spectral densities. *Journal of the Royal Statistical Society, Series B* **24**, 185–198.

Davies, R. B. (1980). ASS 155: The distribution of a linear combination of chi-squared random variables. *Applied Statistics* **29**, 323–333.

Day, N. E. and D. F. Kerridge (1967). A general maximum likelihood discriminant. *Biometrics* **23**, 313–323.

De Forest, E. L. (1873). On some methods of interpolation applicable to the graduation of irregular series, such as tables of mortality. Annual Report of the Board of Regents of the Smithsonian Institution for 1873.

De Forest, E. L. (1877). On adjustment formulas. *The Analyst* **4**, 79–86.

Deng, K. and A. W. Moore (1996). Locally weighted logistic regression for memory-based classification. Technical report, The Robotics Institute, Carnegie-Mellon University.

Devroye, L., L. Györfi, A. Krzyzak and G. Lugosi (1994). On the strong universal consistency of nearest neighbor regression function estimates. *The Annals of Statistics* **22**, 1371–1385.

Dickey, J. M. (1968). Smooth estimates for multinomial cell probabilities. *The Annals of Mathematical Statistics* **39**, 561–566.

Dierckx, P. (1993). *Curve and Surface Fitting with Splines*. Oxford: Clarendon Press.

Donoho, D. L. and I. M. Johnstone (1994). Ideal spatial adaptation by wavelet shrinkage. *Biometrika* **81**, 425–455.

Draper, N. R. and H. Smith (1981). *Applied Regression Analysis* (Second ed.). New York: John Wiley & Sons.

Droge, B. (1996). Some comments on cross-validation. In W. Härdle and M. G. Schimek (Eds.), *Statistical Theory and Computational Aspects of Smoothing*, Heidelberg, pp. 178–199. Physica-Verlag.

Duin, R. P. W. (1976). On the choice of smoothing parameter for Parzen estimators of probability density functions. *IEEE Transactions on Computing* **C-25**, 1175–1179.

Efron, B. (1975). The efficiency of logistic regression compared to normal discriminant analysis. *Journal of the American Statistical Association* **70**, 892–898.

Eilers, P. H. C. and B. D. Marx (1996). Flexible smoothing with B-splines and penalties (with discussion). *Statistical Science* **11**, 89–121.

Engle, R., C. Granger, J. Rice and A. Weiss (1986). Nonparametric estimates of the relation between weather and electricity sales. *Journal of the American Statistical Association* **81**, 310–320.

Epanechnikov, V. K. (1969). Nonparametric estimation of a multivariate probability density. Теория Вероятностей и ее Применения *(Theory of Probability and its Applications)* **14**, 156–162 (153–158).

Eubank, R. L. and P. L. Speckman (1993a). A bias reduction theorem with applications in nonparametric regression. *Scandinavian Journal of Statistics* **18**, 211–222.

Eubank, R. L. and P. L. Speckman (1993b). Confidence bands in nonparametric regression. *Journal of the American Statistical Association* **88**, 1287–1301.

Fan, J. (1993). Local linear regression smoothers and their minimax efficiencies. *The Annals of Statistics* **21**, 196–216.

Fan, J. and I. Gijbels (1994). Censored regression: Local linear approximations and their applications. *Journal of the American Statistical Association* **89**, 560–570.

Fan, J. and I. Gijbels (1995a). Adaptive order polynomial fitting: bandwidth robustification and bias reduction. *Journal of Computational and Graphical Statistics* **4**, 213–227.

Fan, J. and I. Gijbels (1995b). Data-driven bandwidth selection in local polynomial fitting: variable bandwidth and spatial adaptation. *Journal of the Royal Statistical Society, Series B* **57**, 371–394.

Fan, J. and I. Gijbels (1996). *Local Polynomial Modelling and its Applications*. London: Chapman and Hall.

Fan, J., N. E. Heckman and M. P. Wand (1995). Local polynomial kernel regression for generalized linear models and quasi-likelihood functions. *Journal of the American Statistical Association* **90**, 141–150.

Faraway, J. and J. Sun (1995). Simultaneous confidence bands for linear regression with heteroscedastic errors. *Journal of the American Statistical Association* **90**, 1094–1098.

Farrell, R. H. (1972). On the best obtainable asymptotic rates of convergence in estimation of a density function at a point. *The Annals of Mathematical Statistics* **43**, 170–180.

Firth, D., J. Glosup and D. V. Hinkley (1991). Model checking with nonparametric curves. *Biometrika* **78**, 245–252.

Fisher, N. I. (1993). *Statistical Analysis of Circular Data*. Cambridge: Cambridge University Press.

Fisher, N. I. and A. J. Lee (1992). Regression models for angular response. *Biometrics* **48**, 665–677.

Fisher, R. A. (1936). The use of multiple measurements in taxonomic problems. *Annals of Eugenics* **7, Part II**, 179–188.

Fix, E. and J. L. J. Hodges (1951). Nonparametric discrimination: Consistency properties. Technical report, USAF School of Aviation Medicine.

Friedman, J. H. (1991). Multivariate adaptive regression splines (with discussion). *Annals of Statistics* **19**, 1–141.

Friedman, J. H. (1994). Flexible metric nearest neighbor classification. Technical report, Department of Statistics, Stanford University.

Friedman, J. H. (1997). On bias, variance, 0/1-loss, and the curse of dimensionality. *Data Mining and Knowledge Discovery* **1**, 55–77.

Friedman, J. H., J. L. Bentley and R. A. Finkel (1977). An algorithm for finding best matches in logarithmic expected time. *ACM Transactions on Mathematical Software* **3**, 209–226.

Friedman, J. H. and W. Stuetzle (1981). Projection pursuit regression. *Journal of the American Statistical Association* **76**, 817–823.

Friedman, J. H. and W. Stuetzle (1982). Smoothing of scatterplots. Technical Report Orion 3, Department of Statistics, Stanford University.

Gasser, T., A. Kneip and W. Köhler (1991). A flexible and fast method for automatic smoothing. *Journal of the American Statistical Association* **86**, 643–652.

Gasser, T., H. G. Müller and V. Mammitzsch (1985). Kernels for nonparametric curve estimation. *Journal of the Royal Statistical Society, Series B* **47**, 238–252.

Gentleman, R. and J. Crowley (1991). Local full likelihood estimation for the proportional hazards model. *Biometrics* **47**, 1283–1296.

Goldenshluger, A. and A. Nemirovski (1997). On spatial adaptive estimation of nonparametric regression. *Mathematical Methods of Statistics* **6**, 135–170.

Gould, A. L. (1969). A regression model for angular variates. *Biometrics* **25**, 683–670.

Gram, J. P. (1879). *Om Raekkeuviklinger, bestendeved Hjaelp af de mindste Kvadraters Methode.* Copenhagen: A. F. Host & Son.

Gray, R. J. (1996). Hazard regression using ordinary nonparametric regression smoothers. *Journal of Computational and Graphical Statistics* **5**, 190–207.

Green, P. J. (1987). Penalized likelihood for general semi-parametric regression models. *International Statistical Review* **55**, 245–260.

Green, P. J. and B. W. Silverman (1994). *Nonparametric Regression and Generalized Linear Models. A Roughness Penalty Approach.* London: Chapman and Hall.

Grenander, U. and M. Rosenblatt (1953). Statistical spectral analysis of time series arising from stationary stochastic processes. *The Annals of Mathematical Statistics* **24**, 537–558.

Grosse, E. (1989). LOESS: Multivariate smoothing by moving least squares. In C. K. Chui, L. L. Schumaker and J. D. Ward (Eds.), *Approximation Theory VI*, Volume II, pp. 373–376. Academic Press.

Gu, C. (1998). Model indexing and smoothing parameter selection in nonparametric regression (with discussion). *Statistica Sinica* **8**(3), 607–646.

Habbema, J. D. F., J. Hermans and K. Van Der Broek (1974). A stepwise discriminant analysis program using density estimation. In G. Bruckman (Ed.), *COMPSTAT 1974, Proceedings in Computational Statistics*, Vienna, pp. 101–110. Physica-Verlag.

Hall, P. and J. S. Marron (1987). Extent to which least-squares cross-validation minimizes integrated squared error in nonparametric density estimation. *Probability Theory and Related Fields* **74**, 567–581.

Hall, P., S. J. Sheather, M. C. Jones and J. S. Marron (1991). On optimal data-based bandwidth selection in kernel density estimation. *Biometrika* **78**, 263–270.

Hall, P. and T. E. Wehrly (1991). A geometrical method for removing edge effects from kernel-type nonparametric regression estimators. *Journal of the American Statistical Association* **86**, 665–672.

Hampel, F. R., E. M. Ronchetti, P. Rousseeuw and W. Stahel (1986). *Robust Statistics: The Approach Based on Influence Functions*. New York: Wiley.

Härdle, W. (1990). *Applied Nonparametric Regression*. Cambridge: Cambridge University Press.

Härdle, W., P. Hall and J. S. Marron (1992). Regression smoothing parameters that are not far from their optimal. *Journal of the American Statistical Association* **87**, 227–233.

Härdle, W. and S. Luckhaus (1984). Uniform consistency of a class of regression function estimators. *The Annals of Statistics* **12**, 612–623.

Härdle, W. and E. Mammen (1993). Comparing nonparametric versus parametric regression fits. *The Annals of Statistics* **21**, 1926–1947.

Hart, J. D. (1997). *Nonparametric Smoothing and Lack-of-Fit Tests*. New York: Springer-Verlag.

Hastie, T. J. (1992). Generalized additive models. In J. M. Chambers and T. J. Hastie (Eds.), *Statistical Models in S*, pp. 249–307. Pacific Grove: Wadsworth and Brooks/Cole.

Hastie, T. J. and C. R. Loader (1993). Local regression: Automatic kernel carpentry (with discussion). *Statistical Science* **8**, 120–143.

Hastie, T. J. and D. Pregibon (1992). Generalized linear models. In J. M. Chambers and T. J. Hastie (Eds.), *Statistical Models in S*, pp. 195–247. Pacific Grove: Wadsworth and Brooks/Cole.

Hastie, T. J. and R. J. Tibshirani (1986). Generalized additive models (with discussion). *Statistical Science* **1**, 297–318.

Hastie, T. J. and R. J. Tibshirani (1990). *Generalized Additive Models*. London: Chapman and Hall.

Hastie, T. J. and R. J. Tibshirani (1992). Varying-coefficient models. *Journal of the Royal Statistical Society, Series B* **55**, 757–796.

Haupt, G. and U. Mansmann (1995). CART for survival data. Statlib Archive.
http://lib.stat.cmu.edu/S/survcart

Henderson, R. (1916). Note on graduation by adjusted average. *Transactions of the Actuarial Society of America* **17**, 43–48.

Henderson, R. (1924a). A new method of graduation. *Transactions of the Actuarial Society of America* **25**, 29–40.

Henderson, R. (1924b). Some points in the general theory of graduation. In *Proceedings of the International Mathematical Congress*, pp. 815–820.

Henderson, R. and H. N. Sheppard (1919). *Graduation of Mortality and Other Tables*. New York: Actuarial Society of America.

Heyde, C. C. (1997). *Quasi-Likelihood and its Application*. New York: Springer.

Hinkley, D. V. (1970). Inference about the change-point in a sequence of random variables. *Biometrika* **57**, 1–17.

Hjellvik, V., Q. Yao and D. Tjøstheim (1996). Linearity testing using local polynomial approximation. Technical Report UKC/IMS/96/19, Institute of Mathematics and Statistics, University of Kent at Canterbury.
ftp://ftp.ukc.ac.uk/pub/maths/reports/1996/19/19.ps.gz

Hjort, N. L. (1993). Dynamic likelihood hazard rate estimation. Technical Report 4, Institute of Mathematics, University of Oslo.

Hjort, N. L. and M. C. Jones (1996). Locally parametric nonparametric density estimation. *The Annals of Statistics* **24**, 1619–1647.

Hoem, J. M. (1983). The reticent trio: Some little-known early discoveries in life insurance mathematics by L. H. F. Oppermann, T. N. Thiel and J. P. Gram. *International Statistical Review* **51**, 213–221.

Horspool, R. N. (1986). *C Programming in the Berkeley Unix Environment*. Scarborough, Ontario: Prentice-Hall.

Horváth, L. (1991). On L_p-norms of multivariate density estimators. *The Annals of Statistics* **19**, 1933–1949.

Hotelling, H. (1939). Tubes and spheres in n-spaces, and a class of statistical problems. *American Journal of Mathematics* **61**, 440–460.

Huber, P. J. (1964). Robust estimation of a location parameter. *The Annals of Mathematical Statistics* **35**, 73–101.

Huber, P. J. (1981). *Robust Statistics*. New York: Wiley.

Ichimura, H. and P. Todd (1999). Implementing nonparametric and semiparametric estimators. In *Handbook of Econometrics*, Volume 5. North Holland.

Ihaka, R. and R. Gentleman (1996). R: A language for data analysis and graphics. *Journal of Computational and Graphical Statistics* **5**, 299–314.

Imhof, J. P. (1961). Computing the distribution of quadratic forms in normal variables. *Biometrika* **48**, 419–426.

Ioffe, M. O. and V. Y. Katkovnik (1989). Pointwise and uniform convergence with probability 1 of nonparametric regression estimates. Автоматика И Телемеханика *(Automation and Remote Control)* **50**(12), 59–68 (1659–1667).

Izenman, A. J. and C. J. Sommer (1988). Philatelic mixtures and multimodal densities. *Journal of the American Statistical Association* **83**, 941–953.

Johnson, R. A. and T. E. Wehrly (1978). Some angular-linear distributions and related regression models. *Journal of the American Statistical Association* **73**, 602–606.

Jones, M. C., J. S. Marron and S. J. Sheather (1996). A brief survey of bandwidth selection for density estimation. *Journal of the American Statistical Association* **91**, 401–407.

Jose, C. T. and B. Ismail (1997). Estimation of jump points in nonparametric regression through residual analysis. *Communications in Statistics - Theory and Methods* **26**, 2583–2607.

Kaplan, E. L. and P. Meier (1958). Nonparametric estimation from incomplete observations. *Journal of the American Statistical Association* **53**, 457–481.

Katkovnik, V. Y. (1979). Linear and nonlinear methods of nonparametric regression analysis. Автоматика *(Soviet Automatic Control)* **5**, 35–46 (25–34).

Katkovnik, V. Y. (1983). Convergence of linear and nonlinear nonparametric estimates of "kernel" type. Автоматика И Телемеханика *(Automation and Remote Control)* **44**(4), 108–120 (495–505).

Katkovnik, V. Y. (1985). Непараметрическая Идентификация И Сглаживание Данных: Метод Локальной Аппроксимации *(Nonparametric Identification and Smoothing of Data: Local Approximation Method)*. Moscow: Nauka.

Katkovnik, V. Y. (1996). Adaptive local polynomial periodogram for time-varying frequency estimation. In *Proceedings of the IEEE-SP International Symposium on Time-Frequency and Time-Scale Analysis (TFTS-96)*, pp. 329–332.

Katkovnik, V. Y. (1998). On multiple window local polynomial approximation with varying adaptive bandwidths. In *COMPSTAT 1998, Proceedings in Computational Statistics*, pp. 353–358. Physica-Verlag.

Katkovnik, V. Y. (1999). A new method for varying adaptive bandwidth selection. *IEEE Transactions on Signal Processing* **47**, To Appear.

Katkovnik, V. Y. and L. J. Stankovic (1998). Periodogram with varying and data-driven window length. *Signal Processing* **67**(3), 000–000.

Kendall, M. and J. K. Ord (1990). *Time Series* (third ed.). New York: Oxford University Press.

Kenny, P. B. and J. Durbin (1982). Local trend estimation and seasonal adjustment of economic and social time series (with discussion). *Journal of the Royal Statistical Society, Series A* **145**, 1–41.

Kharin, Y. S. (1983). Analysis and optimization of Rosenblatt-Parzen classifier with the aid of asymptotic expansions. Автоматика И Телемеханика *(Automation and Remote Control)* **44**(1), 91–100 (72–80).

Khas'minskii, R. Z. (1978). A lower bound on the risks of non-parametric estimates of densities in the uniform metric. Теория Вероятностей и ее Применения *(Theory of Probability and its Applications)* **23**(4), 824–828 (794–798).

Kimber, A. C. and A. R. Hansford (1993). A statistical analysis of batting in cricket. *Journal of the Royal Statistical Society, Series A* **156**, 443–455.

Knafl, G., J. Sacks and D. Ylvisaker (1985). Confidence bands for regression functions. *Journal of the American Statistical Association* **80**, 683–691.

Knowles, M. and D. Siegmund (1989). On Hotelling's geometric approach to testing for a nonlinear parameter in regression. *International Statistical Review* **57**, 205–220.

Konakov, V. D. and V. I. Piterbarg (1984). On the convergence rate of maximal deviation distribution for kernel regression estimates. *Journal of Multivariate Analysis* **15**, 279–294.

Kooperberg, C. and C. J. Stone (1992). Logspline density estimation for censored data. *Journal of Computational and Graphical Statistics* **1**, 301–328.

Kooperberg, C., C. J. Stone and Y. K. Truong (1995). Hazard regression. *Journal of the American Statistical Association* **90**, 78–94.

Koul, H., V. Sursala and J. Van Ryzin (1981). Regression analysis with randomly right-censored data. *The Annals of Statistics* **9**, 1276–1288.

Krause, A. and M. Olson (1997). *The Basics of S and S-PLUS*. New York: Springer-Verlag.

Lai, T. L. and Z. Ying (1991). Large sample theory of a modified Buckley-James estimator for regression analysis with censored data. *The Annals of Statistics* **19**, 1370–1402.

Lancaster, P. and K. Salkaus kas (1981). Surfaces generated by moving least squares methods. *Mathematics of Computation* **37**, 141–158.

Lancaster, P. and K. Salkauskas (1986). *Curve and Surface Fitting: An Introduction*. London: Academic Press.

Langaas, M. (1995). Discrimination and classification. Technical report, Department of Mathematical Sciences, The Norwegian Institute of Technology.
http://www.math.ntnu.no/preprint/statistics/ps/S1-1995.ps

Lee, D. (1989). Discontinuity detection and curve fitting. In C. K. Chui, L. L. Schumaker and J. D. Ward (Eds.), *Approximation Theory VI*, Volume I, pp. 299–302. Academic Press.

Legostaeva, I. L. and A. N. Shiryayev (1971). Minimax weights in a trend detection problem of a random process. Теория Вероятностей и ее Применения *(Theory of Probability and its Applications)* **16**, 344–349.

Lehmann, E. L. (1986). *Testing Statistical Hypothesis* (Second ed.). New York: John Wiley & Sons.

Lejeune, M. (1985). Nonparametric estimation with kernels: Moving polynomial regression. *Revue de Statistiques Appliquees* **33**, 43–67.

Lejeune, M. and P. Sarda (1992). Smooth estimators of distribution and density functions. *Computational Statistics & Data Analysis* **14**, 457–471.

Lepski, O. V., E. Mammen and V. G. Spokoiny (1997). Optimal spatial adaptation to inhomogeneous smoothness: An approach based on kernel estimates with variable bandwidth selectors. *The Annals of Statistics* **25**, 929–947.

Leurgans, S. (1987). Linear models, random censoring and synthetic data. *Biometrika* **74**, 301–309.

Li, X. and N. E. Heckman (1996). Local linear forecasting. Technical Report 167, Department of Statistics, University of British Columbia. http://www.stat.ubc.ca/research/techreports/167.ps

Loader, C. R. (1991). Inference for a hazard rate change point. *Biometrika* **78**, 749–757.

Loader, C. R. (1994). Computing nonparametric function estimates. In *Computing Science and Statistics: Proceedings of the 26th Symposium on the Interface*, pp. 356–361.

Loader, C. R. (1996a). Change point estimation using nonparametric regression. *The Annals of Statistics* **24**, 1667–1678.

Loader, C. R. (1996b). Local likelihood density estimation. *The Annals of Statistics* **24**, 1602–1618.

Loader, C. R. (1999). Bandwidth selection: classical or plug-in? *The Annals of Statistics* **27**, To Appear.

Low, M. G. (1993). Renormalizing upper and lower bounds for integrated risk in the white noise model. *The Annals of Statistics* **21**, 577–589.

Macaulay, F. R. (1931). *Smoothing of Time Series*. New York: National Bureau of Economic Research.

Maechler, M. (1992). Robustifying a (local, nonparametric) regression estimatr. Technical report, Swiss Federal Institute of Technology (ETH), Zurich.

Mallows, C. L. (1973). Some comments on C_p. *Technometrics* **15**, 661–675.

Marron, J. S. (1996). A personal view of smoothing and statistics. In W. Härdle and M. G. Schimek (Eds.), *Statistical Theory and Computational Aspects of Smoothing*, Heidelberg, pp. 1–9. Physica-Verlag.

Marron, J. S. and M. P. Wand (1992). Exact mean integrated squared error. *The Annals of Statistics* **20**, 712–736.

McCullagh, P. and J. A. Nelder (1989). *Generalized Linear Models*. London: Chapman and Hall.

McDonald, J. A. and A. B. Owen (1986). Smoothing with split linear fits. *Technometrics* **28**, 195–208.

McLain, D. H. (1974). Drawing contours from arbitrary data. *Computer Journal* **17**, 318–324.

Miller, R. and J. Halpern (1982). Regression with censored data. *Biometrika* **69**, 521–531.

Miller, R. G. (1981). *Survival Analysis*. New York: John Wiley & Sons.

Müller, H.-G. (1984). Smooth optimum kernel estimators of densities, regression curves and modes. *The Annals of Statistics* **12**, 766–774.

Müller, H. G. (1987). Weighted local regression and kernel methods for nonparametric curve fitting. *Journal of the American Statistical Association* **82**, 231–238.

Müller, H.-G. (1988). *Nonparametric Regression Analysis of Longitudinal Data*. Heidelberg: Springer-Verlag.

Müller, H.-G. (1992). Change-points in nonparametric regression analysis. *The Annals of Statistics* **20**, 737–761.

Müller, H. G. and J.-L. Wang (1994). Hazard rate estimation under random censoring with varying kernels and bandwidths. *Biometrics* **50**, 61–76.

Murphy, B. J. and M. A. Moran (1986). Parametric and kernel density methods in discriminant analysis: Another comparison. In S. C. Choi (Ed.), *Statistical Methods of Discrimination and Classification*. Pergamon Press.

Myers, R. H. (1990). *Classical and Modern Regression with Applications* (second ed.). Boston: PWS-Kent Publishing.

Nadaraya, E. A. (1964). On estimating regression. Теория Вероятностей и ее Применения *(Theory of Probability and its Applications)* **9**, 157–159 (141–142).

Naiman, D. Q. (1990). On volumes of tubular neighborhoods of spherical polyhedra and statistical inference. *The Annals of Statistics* **18**, 685–716.

Nelder, J. A. and D. Pregibon (1987). An extended quasi-likelihood function. *Biometrika* **74**, 221–232.

Opsomer, J. D. and D. Ruppert (1997). Fitting a bivariate additive model by local polynomial regression. *The Annals of Statistics* **25**, 186–211.

Park, B. U. and J. S. Marron (1990). Comparison of data-driven bandwidth selectors. *Journal of the American Statistical Association* **85**, 66–72.

Parzen, E. (1961). Mathematical considerations in the estimation of spectra: Comments on the discussion of Messrs Tukey and Goodman. *Technometrics* **3**, 167–190; 232–234.

Parzen, E. (1962). On estimation of a probability density function and mode. *The Annals of Mathematical Statistics* **33**, 1065–1076.

Prakasa Rao, B. L. S. (1983). *Nonparametric Functional Estimation*. Orlando: Academic Press.

Pregibon, D. (1981). Logistic regression diagnostics. *The Annals of Statistics* **9**, 705–724.

Qiu, P. and B. Yandell (1998). A local polynomial jump-detection algorithm in nonparametric regression. *Technometrics* **40**, 141–152.

Raz, J. (1990). Testing for no effect when estimating a smooth regression function by nonparametric regression. *Journal of the American Statistical Association* **85**, 132–138.

Reaven, G. M. and R. G. Miller (1979). An attempt to define the nature of chemical diabetes using a multidimensional analysis. *Diabetologia* **16**, 17–24.

Rice, J. (1984). Bandwidth choice for nonparametric regression. *The Annals of Statistics* **12**, 1215–1230.

Rice, S. O. (1939). The distribution of the maxima of a random curve. *American Journal of Mathematics* **61**, 409–416.

Rigby, R. A. and M. D. Stasinopoulos (1996). Mean and dispersion additive models. In W. Härdle and M. G. Schimek (Eds.), *Statistical Theory and Computational Aspects of Smoothing*, pp. 215–230. Heidelberg: Physica-Verlag.

Ripley, B. D. (1994). Flexible non-linear approaches to classification. In B. Cherkassky, J. H. Friedman and H. Wechsler (Eds.), *From Statistics to Neural Networks. Theory and Pattern Recognition Applications*, pp. 105–126. Springer-Verlag.

Robinson, P. (1988). Root-n consistent semiparametric regression. *Econometrica* **59**, 1329–1363.

Rosenblatt, M. (1956). Remarks on some nonparametric estimates of a density function. *The Annals of Mathematical Statistics* **27**, 832–837.

Rudemo, M. (1982). Empirical choice of histograms and kernel density estimators. *Scandinavian Journal of Statistics* **9**, 65–78.

Ruppert, D., S. J. Sheather and M. P. Wand (1995). An effective bandwidth selector for local least squares regression. *Journal of the American Statistical Association* **90**, 1257–1270.

Ruppert, D. and M. P. Wand (1994). Multivariate locally weighted least squares regression. *The Annals of Statistics* **22**, 1346–1370.

Ruppert, D., M. P. Wand, U. Holst and O. Hössjer (1997). Local polynomial variance function estimation. *Technometrics* **39**, 262–273.

Sacks, J. and D. Ylvisaker (1978). Linear estimation for approximately linear models. *The Annals of Statistics* **6**, 1122–1137.

Sacks, J. and D. Ylvisaker (1981). Asymptotically optimum kernels for density estimation at a point. *The Annals of Statistics* **9**, 334–346.

Satterthwaite, F. E. (1946). An approximate distribution of estimates of variance components. *Biometrics Bulletin* **2**, 110–114.

Savitzky, A. and M. J. E. Golay (1964). Smoothing and differentiation of data by simplified least squares procedures. *Analytical Chemistry* **36**, 1627–1639.

Scheffé, H. (1959). *The Analysis of Variance*. New York: John Wiley & Sons.

Schiaparelli, G. V. (1866). Sul modo di ricavare la vera espressione delle leggi delta natura dalle curve empiricae. *Effemeridi Astronomiche di Milano per l'Arno* **857**, 3–56.

Schmee, J. and G. J. Hahn (1979). A simple method for regression analysis with censored data (with discussion). *Technometrics* **21**, 417–434.

Schmidt, G., R. Mattern and F. Schüler (1981). Biomechanical investigation to determine physical and traumatalogical differentiation criteria for the maximum load capacity of head and vertebral column with and without protective helmet under effects of impact. Final report phase III, Project 65, Universität Heidelberg.

Schuster, E. F. and G. G. Gregory (1981). On the nonconsistency of maximum likelihood nonparametric density estimators. In *Computing Science and Statistics: Proceedings of the 13th Symposium on the Interface*, pp. 295–298.

Schwarz, G. (1978). Estimating the dimension of a model. *The Annals of Statistics* **6**, 461–464.

Scott, D. W. (1992). *Multivariate Density Estimation: Theory, Practice and Visualization.* New York: John Wiley & Sons.

Scott, D. W., R. A. Tapia and J. R. Thompson (1977). Kernel density estimation revisited. *Journal of Nonlinear Analysis: Theory, Methods and Applications* **1**, 339–372.

Scott, D. W. and G. R. Terrell (1987). Biased and unbiased cross-validation in density estimation. *Journal of the American Statistical Association* **82**, 1131–1146.

Seal, H. L. (1981). Graduation by piecewise cubic polynomials: a historical review. *Blätter. Deutshe Gesellschaft für Versicherungsmathematik* **14**, 237–253.

Seidel, H. (1997). Functional data fitting and fairing with triangular B-splines. In A. Le Méhauté, C. Rabut and L. L. Schumaker (Eds.), *Surface Fitting and Multiresolution Methods.* Nashville: Vanderbilt University Press.

Seifert, B. and T. Gasser (1996). Finite-sample variance of local polynomials: analysis and solutions. *Journal of the American Statistical Association* **91**, 267–275.

Sergeev, V. L. (1979). Use of estimates of local approximation of probability density. Автоматика И Телемеханика *(Automation and Remote Control)* **40**(7), 56–61 (971–995).

Severini, T. A. and J. Staniswalis (1994). Quasi-likelihood estimation in semiparametric models. *Journal of the American Statistical Association* **89**, 501–511.

Sheather, S. J. (1992). The performance of six popular bandwidth selection methods on some real datasets. *Computational Statistics* **7**, 225–250.

Sheather, S. J. and M. C. Jones (1991). A reliable data-based bandwidth selection method for kernel density estimation. *Journal of the Royal Statistical Society, Series B* **53**, 683–690.

Sheppard, W. F. (1914a). Graduation by reduction of mean square error. I. *Journal of the Institute of Actuaries* **48**, 171–185.

Sheppard, W. F. (1914b). Graduation by reduction of mean square error. II. *Journal of the Institute of Actuaries* **48**, 390–412.

Shiryayev, A. N. (1984). *Probability*. New York: Springer-Verlag.

Shiskin, J., A. H. Young and J. C. Musgrave (1967). The X-11 variant of the census method II seasonal adjustment program. Technical paper 15, Bureau of the Census, U. S. Department of Commerce.

Siegmund, D. O. and K. J. Worsley (1995). Testing for a signal with unknown location and scale in a stationary Gaussian random field. *The Annals of Statistics* **23**, 608–639.

Silverman, B. W. (1986). *Density Estimation for Statistics and Data Analysis*. London: Chapman and Hall.

Simonoff, J. S. (1987). Probability estimation via smoothing in sparse contingency tables with ordered categories. *Statistics and Probability Letters* **5**, 55–63.

Simonoff, J. S. (1995). Smoothing categorical data. *Journal of Statistical Planning and Inference* **47**, 41–69.

Simonoff, J. S. (1996). *Smoothing Methods in Statistics*. New York: Springer.

Snee, R. D. (1977). Validation of regression models: Methods and examples. *Technometrics* **19**, 415–428.

Spector, P. (1994). *An Introduction to S and S-PLUS*. Duxbury Press.

Spencer, J. (1904). On the graduation of rates of sickness and mortality. *Journal of the Institute of Actuaries* **38**, 334–343.

Staniswalis, J. G. (1989). On the kernel estimate of a regression function in likelihood based models. *Journal of the American Statistical Association* **84**, 276–283.

Staniswalis, J. G. and T. A. Severini (1991). Diagnostics for assessing regression models. *Journal of the American Statistical Association* **86**, 684–692.

Stigler, S. M. (1978). Mathematical statistics in the early states. *The Annals of Statistics* **6**, 239–265.

Stoker, T. M. (1993). Smoothing bias in density derivative estimation. *Journal of the American Statistical Association* **88**, 855–871.

Stone, C. J. (1977). Consistent nonparametric regression (with discussion). *The Annals of Statistics* **5**, 595–645.

Stone, C. J. (1980). Optimal rates of convergence for nonparametric estimators. *The Annals of Statistics* **8**, 1348–1360.

Stone, C. J. (1982). Optimal global rates of convergence for nonparametric regression. *The Annals of Statistics* **10**, 1040–1053.

Stone, C. J., M. H. Hansen, C. Kooperberg and Y. K. Truong (1997). Polynomial splines and their tensor products in extended linear modeling (with discussion). *The Annals of Statistics* **25**, 1371–1470.

Stone, M. (1974). Cross-validating choice and assessment of statistical predictions (with discussion). *Journal of the Royal Statistical Society, Series B* **36**, 111–47.

Stuetzle, W. and Y. Mittal (1979). Some comments on the asymptotic behavior of robust smoothers. In T. Gasser and M. Rosenblatt (Eds.), *Smoothing Techniques for Curve Estimation*, pp. 191–195. Springer-Verlag.

Sun, J. (1993). Tail probabilities of the maxima of Gaussian random fields. *The Annals of Probability* **21**, 34–71.

Sun, J. and C. R. Loader (1994). Simultaneous confidence bands in linear regression and smoothing. *The Annals of Statistics* **22**, 1328–1345.

Taylor, C. C. (1989). Bootstrap choice of the smoothing parameter in kernel density estimation. *Biometrika* **76**, 705–712.

Tibshirani, R. J. (1984). *Local Likelihood Estimation*. Ph. D. thesis, Department of Statistics, Stanford University.

Tibshirani, R. J. and T. J. Hastie (1987). Local likelihood estimation. *Journal of the American Statistical Association* **82**, 559–567.

Titterington, D. M. (1980). A comparative study of kernel-based density estimates for categorical data. *Technometrics* **22**, 259–268.

Tsybakov, A. B. (1982). Robust estimation of a function. Проблемы Передачи Информации *(Problems of Information Transmission)* **18**(3), 39–52 (190–201).

Tsybakov, A. B. (1986). Robust reconstruction of functions by the local-approximation method. Проблемы Передачи Информации *(Problems of Information Transmission)* **22**, 69–84 (133–146).

Tsybakov, A. B. (1987). On the choice of bandwidth in kernel nonparametric regression. Теория Вероятностей и ее Применения *(Theory of Probability and its Applications)* **32**(1), 142–147.

Van Ness, J. and C. Simpson (1976). On the effects of dimension reduction in discriminant analysis. *Technometrics* **18**, 175–187.

Venables, W. N. and B. D. Ripley (1997). *Modern Applied Statistics with S-Plus* (Second ed.). New York: Springer.

Volf, P. (1989). A nonparametric analysis of proportional hazard regression model. Проблемы Управления и Теории Информации *(Problems of Control and Information Theory)* **18**(5), 311–322.

Wahba, G. (1990). *Spline Models for Observational Data*. Philadelphia: SIAM.

Wahba, G. and S. Wold (1975). A completely automatic French curve. *Communications in Statistics* **4**, 1–17.

Wallis, K. F. (1974). Seasonal adjustment and relations between variables. *Journal of the American Statistical Association* **69**, 18–31.

Wand, M. P. and M. C. Jones (1995). *Kernel Smoothing*. London: Chapman and Hall.

Wang, F. T. and D. W. Scott (1994). The L_1 method for robust nonparametric regression. *Journal of the American Statistical Association* **89**, 65–76.

Wang, Z., T. Isaksson and B. R. Kowalski (1994). New approach for distance measurement in locally weighted regression. *Analytical Chemistry* **66**, 249–260.

Watson, G. and M. R. Leadbetter (1964). Hazard analysis. I. *Biometrika* **51**, 175–184.

Watson, G. S. (1964). Smooth regression analysis. *Sankhya Series A* **26**, 359–372.

Wedderburn, R. W. M. (1974). Quasilikelihood functions, generalized linear models and the Gauss-Newton method. *Biometrika* **61**, 439–447.

Weisberg, S. (1985). *Applied Linear Regression Analysis* (Second ed.). New York: John Wiley & Sons.

Weyl, H. (1939). On the volume of tubes. *American Journal of Mathematics* **61**, 461–472.

Whitaker, E. T. (1923). On a new method of graduation. *Proceedings of the Edinburgh Mathematical Society* **41**, 62–75.

Whittle, P. (1958). On the smoothing of probability density functions. *Journal of the Royal Statistical Society, Series B* **20**, 334–343.

Wilk, M. B. and R. Gnanadesikan (1968). Probability plotting methods for the analysis of data. *Biometrika* **55**, 1–17.

Woodroofe, M. (1970). On choosing a delta sequence. *The Annals of Mathematical Statistics* **41**, 1665–1671.

Woolhouse, W. S. B. (1870). Explanation of a new method of adjusting mortality tables, with some observations upon Mr. Makeham's modification of Gompertz's theory. *Journal of the Institute of Actuaries* **15**, 389–410.

Wu, L. and N. B. Tuma (1990). Local hazard models. *Sociological Methodology* **20**, 141–180.

Ye, J. (1998). On measuring and correcting the effects of data mining and model selection. *Journal of the American Statistical Association* **93**, 120–131.

Index